HANDBOOK OF MODERN ELECTRICAL WIRING

HANDBOOK OF MODERN ELECTRICAL WIRING

JOHN E. TRAISTER

RESTON PUBLISHING COMPANY, INC.
A PRENTICE-HALL COMPANY
RESTON, VIRGINIA

Library of Congress Cataloging in Publication Data

Traister, John E
  Handbook of modern electrical wiring.
  Includes index.
  1. Electric wiring—Handbooks, manuals, etc. I. Title.
TK3201.T72   621.319'24   78-27289
ISBN 0-8359-2754-7

Printed in the United States of America

DEDICATION

To my two daughters,
Bonnie and Wendy

CONTENTS

PREFACE

The electrical construction industry in the United States and other nations continues to grow at a phenomenal rate. In the United States alone electrical construction will approach thirty billion dollars a year by the time this book is published. With the continued demand for electrical systems in buildings of all kinds, the need for craftsmen, technicians, engineers, and others in the electrical field increases also.

This publication is intended to provide those entering the electrical construction field with up-to-date guides which cover theory, design, and practical applications to help them approach their work with more confidence. This book will be helpful to all persons working in the electrical construction industry, including vo-tech students, apprentice, electricians, engineers, draftsmen, architects, electrical contractors, and specification writers.

Although this book is mainly for the professional educator and student, it will also serve as a reference book for the professional.

A deep bow is due to all the manufacturers and others who contributed illustrations and reference material for use in this book.

HANDBOOK OF MODERN ELECTRICAL WIRING

1

ELECTRICAL DESIGN CONSTRUCTION AND THE NEC

This chapter introduces the reader to essentials necessary for proper design and installation of electrical systems in buildings. Familiarity with fundamental design procedures, electrical drawings and specifications, and the National Electrical Code (NEC) is necessary for persons working in the electrical industry.

1-1 BUILDING CONSTRUCTION

In the construction of a new building, an architectural firm is usually commissioned to prepare complete *working drawings* and *specifications* for the building. These *construction documents* are then used for estimating and bidding purposes as well as for construction guidelines to be followed during the construction of the building.

If the electrical system for a building is very extensive and complex, the architect usually hires a consulting engineering firm to design and layout the electrical system. The electrical designer also acts as liaison between the architect and the electrical contractor—handling all the details of the electrical construction from the time the design and layout of the work is started, through the bidding and construction sequences, to the final approval and acceptance of the finished electrical installation.

On smaller projects, like residential construction, the architect usually prepares a (fairly standard) layout of the electrical system directly on the architectural drawings. This layout usually indicates location of the lighting outlets and switching arrangement, and perhaps a few duplex receptacles. The details of wiring, number of circuits, size of electrical service, etc., are usually left to the electrical contractor installing the system. For this reason, fundamental electrical design procedures, as stated in the NEC, are important to all electrical workers—not only to design engineers.

A contract for the construction of a building is awarded to a building contractor, normally called the "General Contractor." This contractor, in turn, hires various subcontractors to handle the specialty work, which includes the plumbing, heating and cooling, and the electrical work. These subcontractors are responsible for furnishing all labor and necessary materials to complete their portion of the building contract; they are also responsible for the completion and proper performance of their phase of the work.

1–2   DRAWINGS AND SPECIFICATIONS

The construction documents supplied by an architectural firm for a new building generally include all architectural drawings which show the design and building construction details; these include floor plan layouts, vertical elevations of all building exteriors, and various cross sections of the building. While the number of such drawings varies from job to job depending on its size and complexity, the drawings almost always fall into five general groups:

1   *Site Work*—Site plans include the location of the building on the property and show the location and routing of all outside utilities (water, gas, electricity, sewer, etc.). These serve the building itself, as well as other points within established property lines. Topography lines are sometimes included on the site plan, especially when the building site is on a slope.

2   *Architectural*—These drawings normally include: elevations of all exterior faces of the building; floor plans showing walls, doors, windows, and partitions on each floor; and sufficient cross sections to indicate clearly the various floor levels and details of the foundation, walls, floors, ceilings, and roof construction. Large-scale detail drawings may also be included.

3   *Structural*—Structural drawings are usually included for reinforced-concrete and structural-steel buildings. These drawings are usually prepared by structural consulting engineers.

4   *Mechanical*—The mechanical drawings cover the complete design and layout of the plumbing, piping, heating, ventilating, and air-conditioning systems and related mechanical construction. Electrical-control wiring diagrams for the heating and cooling systems often are included on the mechanical drawings.

5   *Electrical*—The electrical drawings indicate the complete design and layout of the electrical wiring for lighting, power, signals and communications, special electrical systems, and related electrical equipment. These drawings sometimes include: a site

plan showing the location of the building on the property and the interconnecting electrical systems; floor plans showing the location of power outlets, lighting fixtures, panelboards, etc.; power-riser diagrams; a symbol list; schedules; schematic diagrams; and large-scale details where necessary.

In order to be able to ''read'' any of these drawings, one must first become familiar with the meaning of the various symbols, lines, and abbreviations used on the drawings and learn how to interpret the message conveyed by each.

Electrical Drawings

The three types of electrical drawings, which are of primary concern, are:

1   Electrical construction drawings.

2   Single-line block diagrams.

3   Schematic wiring diagrams.

Electrical construction drawings show the physical arrangement and views of specific electrical equipment. These drawings give all the details necessary to construct the electrical installation, such as the plan views and elevation views. For example, Figure 1–1 is a *pictorial sketch* of an electrical main distribution *panelboard housing*. One side of the housing is labeled ''top,'' and others are labeled ''side'' and ''front.''

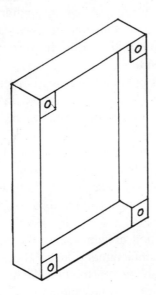

Figure. 1–1   PICTORIAL SKETCH OF AN ELECTRICAL PANEL HOUSING.

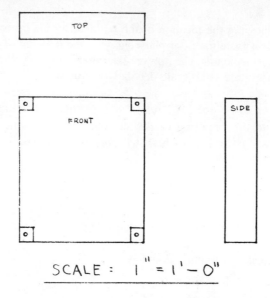

SCALE : 1″ = 1′ – 0″

Figure 1–2   SAME HOUSING IN FIGURE 1–1 REPRESENTED IN ANOTHER FORM.

This same panelboard cabinet is represented in another form in Figure 1–2. The drawing labeled ''top'' is what is seen when the panelboard is viewed directly from above; the drawing labeled ''side'' is the view from the side; the drawing labeled ''front'' shows what the panelboard looks like from the front.

The width of the housing is shown by the horizontal lines in the top view and the front view. The height is shown by the vertical lines in both the front and the side views, and the depth is shown by the vertical lines in the top view and the horizontal lines in the side view.

The three drawings in Figure 1–2 clearly present the shape of the panelboard housing, but the drawings alone give no indiciation of the size of the housing. There are two common methods of indicating the actual length, width, and height of the housing. One is to draw all of the views to some given scale, such as 1-1/2″ = 1′-0″. This means that 1-1/2 inches on the drawing represents 1 foot in the actual construction of the housing. The second method is to give dimensions on the drawings, as is done in Figure 1–3. Note that the gauge and type of material are also given in this drawing—enough data to show clearly how the panelboard housing is to be constructed.

Electrical construction drawings, like the ones just discussed, are used by electrical-equipment manufacturers. Electrical power companies use drawings, as shown in Figure 1–4, giving construction details of a high-voltage transmission line. However, the electrical worker more often sees electrical construction drawings like the one shown in Figure 1–5. This type of construction drawing is normally used to supplement building electrical-system drawings for a special installation and is often referred to as an electrical detail drawing.

Electrical diagrams are drawings that are intended to show, in diagram form, electrical components and their related connections. In diagrams, electrical *symbols* are used extensively to represent the various components. Lines, indicating the size, type, and number of wires that are necessary to complete the electrical circuit, are used to connect these symbols.

The electrical worker often comes into contact with *single-line block diagrams,* which are used extensively by consulting engineering firms to indicate the arrangement of electrical services on electrical working drawings. The power-riser diagram, shown in Figure 1–6, is typical of such drawings. This particular drawing shows all of the panelboards and related equipment, as well as the connecting lines that indicate the circuits and feeders. Notes are used to identify each piece of equipment and to indicate the size of conduit necessary for each circuit or feeder and the number, size, and type of insulation on the conductors in each conduit.

A *schematic wiring diagram* (Figure 1–7) is similar to a single-line block diagram, but the schematic diagram gives more detailed information and shows the actual size and number of wires used for the electrical connections.

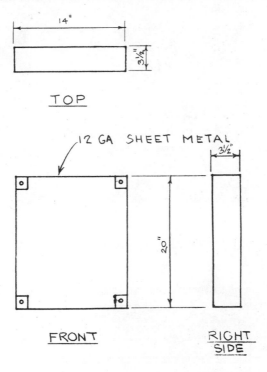

Figure 1–3 METHOD OF SHOWING DIMENSIONS ON DRAWING.

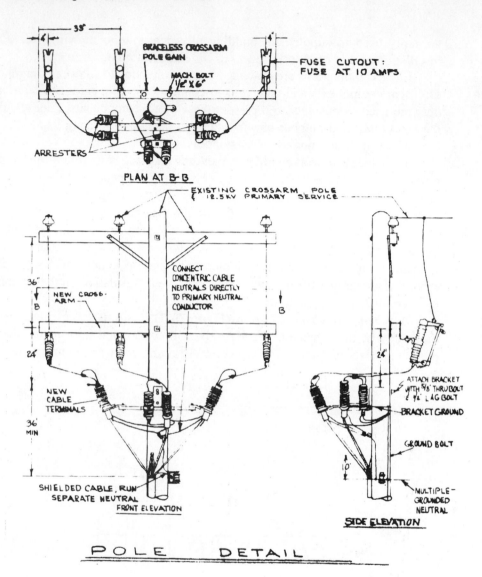

POLE DETAIL

Figure 1–4 ELECTRICAL DRAWING SHOWING DETAILS OF A HIGH-VOLTAGE LINE.

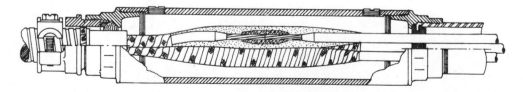

Figure 1–5 ELECTRICAL DETAIL DRAWINGS SHOWING DETAILS OF A CABLE SPLICE.

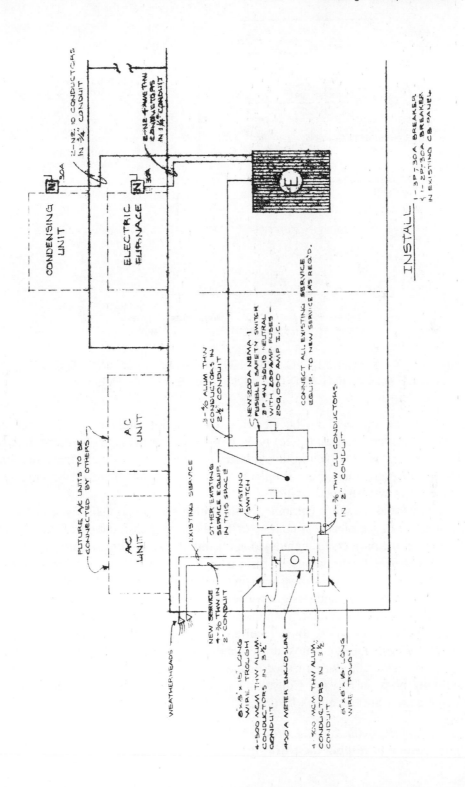

Figure 1-6　POWER RISER DIAGRAM SHOWING PANELBOARDS AND RELATED EQUIPMENT.

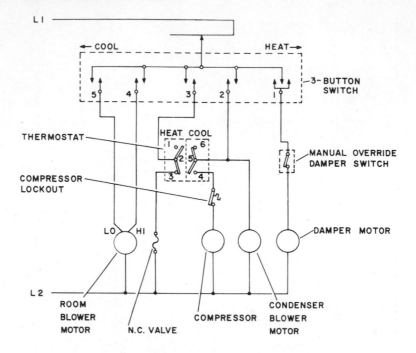

Figure 1–7   A SCHEMATIC WIRING DIAGRAM SHOWING THE CONNEC-
TIONS OF A HEATING AND COOLING SYSTEM CONTROL.

Anyone involved in the electrical construction industry, in any capacity, frequently encounters all (three) types of electrical drawings. Therefore, it is very important for all involved in this industry to understand fully electrical drawings, wiring diagrams, and other supplementary information found in working drawings and in written specifications.[1]

Electrical drawings combined with written specifications show exactly what is required of the electrical contractor for the proper installation of the electrical system on the project. In such drawings, symbols are used to simplify the work of those preparing the drawings. In turn, a knowledge of electrical symbols must be acquired by anyone who must interpret and work from the drawings.

Most engineers and designers use electrical symbols adopted by the United States of America Standards Institute (USASI). However, many of these symbols are frequently modified to suit certain needs for which there is no standard symbol. For this reason, most drawings include a *symbol list* or *legend* as part of the drawings or the written specifications. Figure 1–8 shows a list of electrical symbols prepared by the Consulting Engineers Council/United States and the Construction Specifications Institute, Inc. This list represents a good set of elec-

[1]*Electrical Blueprint Reading,* **Howard W. Sams & Co. (21181) provides a thorough introduction to reading these drawings.**

ELECTRICAL SYMBOLS

SWITCH OUTLETS

Single-Pole Switch	S
Double-Pole Switch	S_2
Three-Way Switch	S_3
Four-Way Switch	S_4
Key-Operated Switch	S_K
Switch and Fusestat Holder	S_FH
Switch and Pilot Lamp	S_P
Fan Switch	S_F
Switch for Low-Voltage Switching System	S_L
Master Switch for Low-Voltage Switching System	S_{LM}
Switch and Single Receptacle	⊖S
Switch and Duplex Receptacle	⊜S
Door Switch	S_D
Time Switch	S_T
Momentary Contact Switch	S_{MC}
Ceiling Pull Switch	Ⓢ
"Hand-Off-Auto" Control Switch	HOA
Multi-Speed Control Switch	M
Push Button	•

RECEPTACLE OUTLETS

Where weather proof, explosion proof, or other specific types of devices are to be required, use the upper-case subscript letters. For example, weather proof single or duplex receptacles would have the uppercase WP subscript letters noted alongside of the symbol. All outlets should be grounded.

Single Receptacle Outlet

Duplex Receptacle Outlet

Triplex Receptacle Outlet

Quadruplex Receptacle Outlet

Duplex Receptacle Outlet-Split Wired

Triplex Receptacle Outlet-Split Wired

250 Volt Receptable Single Phase Use Subscript Letter to Indicate Function (DW-Dishwasher; RA-Range, CD - Clothes Dryer) or numeral (with explanation in symbol schedule)

250 Volt Receptacle Three Phase

Clock Receptacle Ⓒ

Fan Receptacle Ⓕ

Floor Single Receptacle Outlet

Floor Duplex Receptacle Outlet

Floor Special-Purpose Outlet *

Floor Telephone Outlet - Public

Floor Telephone Outlet - Private

Example of the use of several floor outlet symbols to identify a 2, 3, or more gang floor outlet:

Underfloor Duct and Junction Box for Triple, Double or Single Duct System as indicated by the number of parallel lines.

Example of use of various symbols to identify location of different types of outlets or connections for underfloor duct or cellular floor systems:

Cellular Floor Header Duct

*Use numeral keyed to explanation in drawing list of symbols to indicate usage.

Figure 1–8   ELECTRICAL SYMBOL LIST PREPARED BY THE CONSULTING ENGINEERS COUNCIL/U.S. AND THE CONSTRUCTION SPECIFICATIONS INSTITUTE, INC.

CIRCUITING

Wiring Exposed (not in conduit)   —— E ——

Wiring Concealed In Ceiling
  or Wall

Wiring Concealed in Floor   — — — —

Wiring Existing*   - - - - - - - -

Wiring Turned Up   ——————o

Wiring Turned Down   ——————●

Branch Circuit Home Run to   2 1
  Panel Board.   ——→

Number of arrows indicates number of circuits. (A number at each arrow may be used to identify circuit number.)**

BUS DUCTS AND WIREWAYS

Trolley Duct*** | T | | T |

Busway (Service, Feeder, or
  (Plug-in)*** | B | | B |

Cable Trough Ladder or
  Channel*** | C | | C |

Wireway*** | W | | W |

PANELBOARDS, SWITCHBOARDS AND RELATED EQUIPMENT

Flush Mounted Panelboard
  and Cabinet***

Surface Mounted Panelboard
  and Cabinet***

Switchboard, Power Control
  Center, Unit Substations
  (Should be drawn to scale)***

Flush Mounted Terminal Cabinet
  (In small scale drawings the
  TC may be indicated alongside
  the symbol)*** | TC |

Surface Mounted Terminal Cabinet
  (In small scale drawings the
  TC may be indicated alongside
  the symbol)*** | TC |

Pull Box (Identify in relation to
  Wiring System Section and Size)

Motor or Other Power Controller
  (May be a starter or contactor)***

Externally Operated Disconnection
  Switch***

Combination Controller and Disconnection Means***

POWER EQUIPMENT

Electric Motor (HP as indicated)   (1/4)

Power Transformer

Pothead (Cable Termination)

Circuit Element,
  e.g., Circuit Breaker | CB |

Circuit Breaker

Fusible Element

Single-Throw Knife Switch

Double-Throw Knife Switch

Ground   —||ı

Battery   —|—

Contactor | C |

Photoelectric Cell | PE |

Voltage Cycles, Phase   Ex: 480/60/3

Relay | R |

Equipment (Connection (as noted)   ▲

*Note: Use heavy-weight line to identify service and feeders. Indicate empty conduit by notation CO (conduit only).
**Note: Any circuit without further identification indicates two-wire circuit. For a greater number of wires, indicate with cross lines, e.g.:

—|||— 3 wires;   —||||— 4 wires, etc.

Neutral wire may be shown longer. Unless indicated otherwise, the wire size of the circuit is the minimum size required by the specification. Identify different functions, of wiring system. e.g., signalling system by notation or other means.
***Identify by Notation or Schedule

Figure 1–8   (Continued)

REMOTE CONTROL STATIONS FOR MOTORS OR OTHER EQUIPMENT

Pushbutton Station	PB
Float Switch - Mechanical	F
Limit Switch - Mechanical	L
Pneumatic Switch - Mechanical	P
Electric Eye - Beam Source	
Electric Eye - Relay	
Temperature Control Relay Connection (3 Denotes Quantity.)	R₃
Solenoid Control Valve Connection	S
Pressure Switch Connection	P
Aquastat Connection	A
Vacuum Switch Connection	V
Gas Solenoid Valve Connection	G
Flow Switch Connection	F
Timer Connection	T
Limit Switch Connection	L

LIGHTING

	Ceiling Type	Wall Switch
Surface or Pendant Incandescent Fixture (PC = pull chain)	Watts	PC circuit
Surface or Pendant Exit Light	⊗	⊗
Blanked Outlet	Ⓑ	Ⓑ
Junction Box	Ⓙ	Ⓙ
Recessed Incandescent Fixtures		
Surface or Pendant Individual Fluorescent Fixture		
Surface or Pendant Continuous-Row Fluorescent Fixture (Letter indicating controlling switch)		A

Fixture No. ← 1
Wattage ← 100

Symbol not needed at each fixture

*Bare-Lamp Fluorescent Strip

ELECTRIC DISTRIBUTION OR LIGHTING SYSTEM, AERIAL

Pole**	○
Street or Parking Lot Light and Bracket**	
Transformer**	△
Primary Circuit**	
Secondary Circuit**	
Down Guy	
Head Guy	
Sidewalk Guy	
Service Weather Head**	

ELECTRIC DISTRIBUTION OR LIGHTING SYSTEM, UNDERGROUND

Manhole**	M
Handhole**	H
Transformer Manhole or Vault**	TM
Transformer Pad**	TP
Underground Direct Burial Cable (Indicate type, size and number of conductors by notation or schedule)	
Underground Duct Line (Indicate type, size, and number of ducts by cross-section identification of each run by notation or schedule. Indicate type, size, and number of conductors by notation or schedule.	
Street Light Standard Fed From Underground Circuit**	

---

*In the case of continuous-row bare-lamp fluorescent strip above an area-wide diffusing means, show each fixture run, using the standard symbol; indicate area of diffusing means and type by light shading and/or drawing notation.
**Identify by Notation or Schedule

Figure 1–8   (Continued)

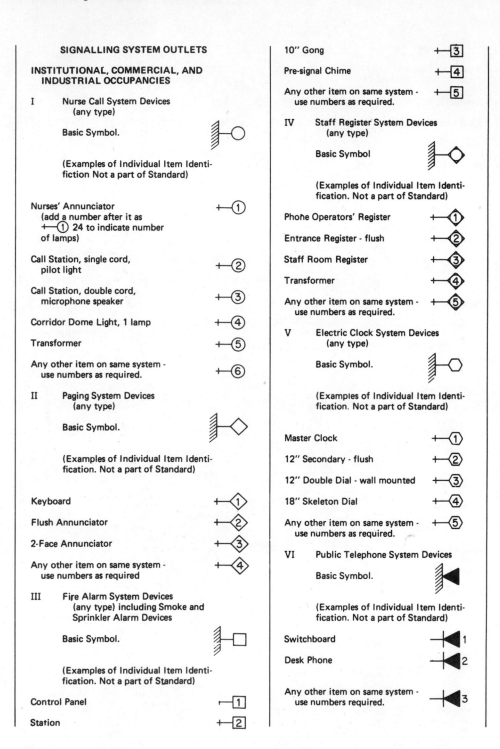

SIGNALLING SYSTEM OUTLETS

INSTITUTIONAL, COMMERCIAL, AND INDUSTRIAL OCCUPANCIES

I   Nurse Call System Devices
     (any type)

     Basic Symbol.

(Examples of Individual Item Identifiction Not a part of Standard)

Nurses' Annunciator
(add a number after it as
⊢①24 to indicate number
of lamps)

Call Station, single cord,
pilot light

Call Station, double cord,
microphone speaker

Corridor Dome Light, 1 lamp

Transformer

Any other item on same system -
use numbers as required.

II   Paging System Devices
      (any type)

      Basic Symbol.

(Examples of Individual Item Identification. Not a part of Standard)

Keyboard

Flush Annunciator

2-Face Annunciator

Any other item on same system -
use numbers as required

III   Fire Alarm System Devices
       (any type) including Smoke and
       Sprinkler Alarm Devices

       Basic Symbol.

(Examples of Individual Item Identification. Not a part of Standard)

Control Panel

Station

10'' Gong

Pre-signal Chime

Any other item on same system -
use numbers as required.

IV   Staff Register System Devices
      (any type)

      Basic Symbol

(Examples of Individual Item Identification. Not a part of Standard)

Phone Operators' Register

Entrance Register - flush

Staff Room Register

Transformer

Any other item on same system -
use numbers as required.

V   Electric Clock System Devices
     (any type)

     Basic Symbol.

(Examples of Individual Item Identification. Not a part of Standard)

Master Clock

12'' Secondary - flush

12'' Double Dial - wall mounted

18'' Skeleton Dial

Any other item on same system -
use numbers as required.

VI   Public Telephone System Devices

      Basic Symbol.

(Examples of Individual Item Identification. Not a part of Standard)

Switchboard

Desk Phone

Any other item on same system -
use numbers required.

Figure 1–8   (Continued)

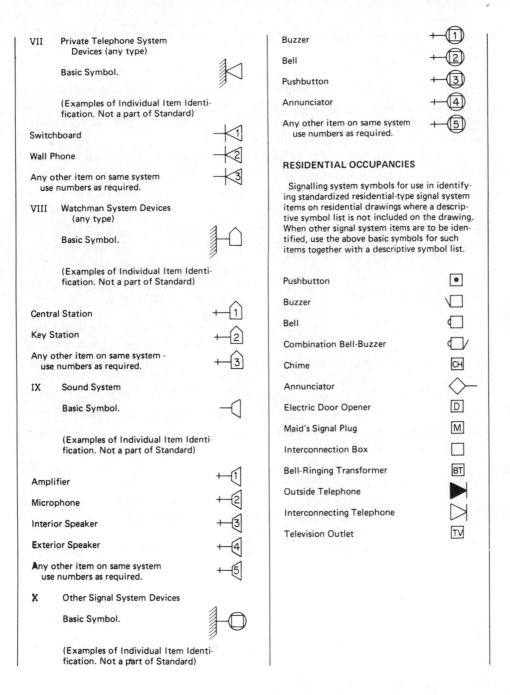

VII   Private Telephone System
         Devices (any type)

         Basic Symbol.

         (Examples of Individual Item Identi-
         fication. Not a part of Standard)

Switchboard

Wall Phone

Any other item on same system
   use numbers as required.

VIII   Watchman System Devices
         (any type)

         Basic Symbol.

         (Examples of Individual Item Identi-
         fication. Not a part of Standard)

Central Station

Key Station

Any other item on same system -
   use numbers as required.

IX   Sound System

         Basic Symbol.

         (Examples of Individual Item Identi-
         fication. Not a part of Standard)

Amplifier

Microphone

Interior Speaker

Exterior Speaker

Any other item on same system
   use numbers as required.

X   Other Signal System Devices

         Basic Symbol.

         (Examples of Individual Item Identi-
         fication. Not a part of Standard)

Buzzer

Bell

Pushbutton

Annunciator

Any other item on same system
   use numbers as required.

RESIDENTIAL OCCUPANCIES

Signalling system symbols for use in identifying standardized residential-type signal system
items on residential drawings where a descriptive symbol list is not included on the drawing.
When other signal system items are to be identified, use the above basic symbols for such
items together with a descriptive symbol list.

Pushbutton

Buzzer

Bell

Combination Bell-Buzzer

Chime

Annunciator

Electric Door Opener

Maid's Signal Plug

Interconnection Box

Bell-Ringing Transformer

Outside Telephone

Interconnecting Telephone

Television Outlet

Figure 1–8   (Continued)

trical symbols because they are, easy to draw, easily interpreted by workmen, and sufficient for most applications.

It is evident from the preceding list that many symbols have the same basic form, but their meanings differ slightly because of the addition of a line, mark, or abbreviation. Therefore, a good procedure to follow in learning the different electrical symbols is first to learn the basic form and then to apply the variations of that form in order to obtain the different meanings.

Note also that some of the symbols listed are abbreviations, such as *XFER* for transfer and *WT* for watertight. Others are simplified *pictographs*, such as Figure 1–9 for an externally operated disconnect switch or Figure 1–10 for a flush-mounted panelboard and housing (cabinet). In some cases, the symbols are combinations of abbreviations and pictographs, such as Figure 1–11 for a flush-mounted terminal cabinet and Figure 1–12 for a nonfusible safety switch.

Electrical Specifications

The *electrical specifications* for a building or project are the written descriptions of the work and duties required of the owner, architect, and engineer. Together with the working drawings, these specifications form the legal basis of the contract requirements for the construction.

Those who use the construction drawings and specifications must always be on the alert for conflicts between the working drawings and the written specifications. Such conflicts frequently occur when:

1   Architects or engineers use standard (proto-type) specifications and attempt to apply them to specific working drawings.

2   Previously prepared standard drawings are changed or amended by reference in the specifications only; the drawings themselves are not changed.

In such situations, the conflicts must be cleared up, preferably *before* the work is undertaken, in order to avoid added cost either to the owner or the electrical contractor.

In general, electrical specifications show the grade of materials to be used and the manner in which the electrical system shall be installed. A sample page from an electrical specification is depicted in Figure 1–13.

1–3.   THE NATIONAL ELECTRICAL CODE (NEC)

Because potential fire and explosion hazards are caused by the improper handling and installation of electrical wiring, certain rules in the selection of materials, quality of workmanship, and precautions for safety must be followed. In order to standardize and simplify these rules, as well as to provide some

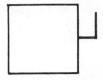

Figure 1–9   SYMBOL FOR EXTERNALLY
OPERATED DISCONNECT SWITCH.

Figure 1–10   SYMBOL FOR FLUSH-MOUNTED
PANELBOARD AND CABINET.

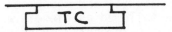

Figure 1–11   SYMBOL FOR FLUSH-MOUNTED
TERMINAL CABINET.

Figure 1–12   SYMBOL FOR A NONFUSIBLE
SAFETY SWITCH.

reliable guide for electrical construction, the National Electrical Code (NEC) was developed. The NEC, originally prepared in 1897, is frequently revised to meet changing conditions, improved equipment and materials, and new fire hazards. It is the product of the best efforts of electrical engineers, manufacturers of electrical equipment, insurance underwriters, firefighters, and other concerned experts throughout the country.

The NEC is now published by the National Fire Protection Association (NFPA).[2] It contains specific rules and regulations intended to help in "the practical safe-guarding of persons and of buildings and their contents from the electrical hazards arising from the use of electricity for light, power, heat. . . ."

Although the NEC itself states, "This Code is to be regarded neither as a design specification nor an instruction manual for untrained persons," it does

[2]NFPA, 60 Batterymarch St., Boston, MA 02138.

DIVISION 16 ELECTRICAL

SECTION 16010 GENERAL PROVISIONS

1. GENERAL:

(A) The "Instructions to Bidders", "General Conditions", "Supplementary General Conditions" and "Special Conditions of the architectural specifications govern work under this Section.

(B) It is understood and agreed that the Electrical Contractor has, by careful examination of the Plans and Specifications, and the site, where appropriate, has satisfied himself as to the nature and location of the work, and all conditions which must be met in order to carry out the work under this Section of the Contract

(C) The Drawings are diagrammatic and indicate generally the locations of material and equipment. These drawings shall be followed as closely as possible. The Electrical Contractor shall coordinate the work under this Section with the architectural, structural, plumbing, heating, and air conditioning, and the drawings of other trades for exact dimensions, clearances and roughing-in locations. This Contractor shall cooperate with all other trades in order to make minor field adjustments to accommodate the work of others.

(D) The Drawings and Specifications are complementary, each to the other, and the work required by either shall be included in the Contract as if called for by both.

(E) The work under this Section includes the furnishing of all labor, materials, equipment and incidentals necessary for the installation of the complete electrical system as shown on the Drawings and as specified.

(F) All work under this Section shall conform to, in all respects, the National Electrical Code and the local electrical code as minimum requirements. Where Plans and Specifications indicate work in excess of the above minimum requirements, the Plans and Specifications shall be followed

16010 Page 1

Figure 1-13   A SAMPLE PAGE FROM AN ELECTRICAL SPECIFICATION.

provide a sound basis for the study of electrical design and installation procedures—under the proper guidance.

The probable reason for this self-analysis is that the NEC further states, "The provisions of this Code constitute a *minimum* standard. Compliance therewith, and proper maintenance, will result in an installation reasonably free from hazard but not necessarily efficient or convenient. . . . Good electrical service and satisfactory results will often require *larger* sizes of wire (in the case of a large voltage drop), *more* branch circuits (to accommodate the ever-increasing number of portable electrical appliances), and better types of equipment than the *minimum* which is here specified."

The NEC, however, has become the "Bible" of the electrical construction industry and anyone involved in electrical work should obtain the latest copy of it, keep it handy at all times, and refer to it frequently.

1–4   LOCAL CODES AND ORDINANCES

A number of towns and cities have their own local electrical code or ordinance. In general, these are based on or are similar to the NEC, but on certain classes of work they may have a few specific rules that are more rigid than the NEC.

Local power companies, may have their own special rules, in addition to the NEC and local ordinances of certain cities, regarding the location of service-entrance wires, meter connections, and similar details; these must be satisfied before connection to a building can be made. Therefore, before design or installation of an electrical system is begun, it is wise to become familiar with all local codes and ordinances.

1–5   UNDERWRITERS' LABORATORIES (UL)

If approved wiring methods are used to install the electrical wiring in a building but low-quality materials are used, the complete installation may still be dangerous. Insurance companies have led the way to require minimum standards of quality in electrical materials, which—through experiment and experience—lead to a maximum of usefulness with the least amount of danger.

The American Insurance Association (AIA) has established a testing organization known as Underwriters' Laboratories (UL), Inc., with testing facilities in several locations throughout the United States. Manufacturers may submit samples of their products to these laboratories for testing before they are manufactured on a large scale. If the product(s) pass the exhaustive tests in accordance with established standards, they are listed in the UL official published list and are then known as "Listed by Underwriters' Laboratories, Inc." Such UL-approved items of materials usually have a *UL label* attached directly to the product. In some cases, this UL label is molded or stamped into the merchandise, such as bakelite, steel, or porcelain parts of wiring devices.

Underwriters' Laboratories is a nonprofit organization supported by the manufacturers who submit merchandise; that is, a fee charged for testing the merchandise pays for the inspectors' expenses and supports the laboratories in general.

A UL label assures the user that the manufacturer of the item submitted samples to the laboratories for testing and that these samples were found to meet the required minimum safety standards. It is presumed that the manufacturer will maintain the same quality in future production of the same item.

Underwriters' Laboratories approval, however, does not mean that the item is approved for all uses. Rather, a UL label means that the item or device as labeled is safe *only* for the purpose for which it was intended; NEC regulations must still be followed. For example, type NM cable cannot be used as a service-entrance cable just because it happens to bear the UL label.

The UL label means that the item meets the *minimum* safety requirements: one item may barely meet the requirements, while another may far exceed them. Therefore, quality cannot really be determined from the UL label alone.

1–6   INSPECTION OF ELECTRICAL WORK

In nearly all but the very small installations, contract specifications require periodic (interim) inspection or approval of the various items of material and the installation methods by the architect or the architect's consulting engineer, as the work progresses. Such inspections are sometimes far and few between, allowing the electricians to perform quite a lot of work prior to inspection. If there is any doubt whether the material and equipment or the method of installation will be accepted by the architect/engineer, an interim inspection should be requested *before* the job has progressed to the point where any required changes would be more costly than necessary.

All electrical construction is usually subject to city, county, state, or power-company inspection to determine whether materials and equipment or installation methods conform to requirements of the NEC and local ordinances. Such inspections are made during the process of roughing-in, at times determined by the extent to which the wiring system is to be concealed within the building structure, and finally upon completion of the job.

It is not unusual to find two or more public authorities having inspection jurisdiction over a given job. Occasionally, inspectors may not agree in their interpretations of the NEC requirements. It is the responsibility of the electrical contractor to have full knowledge of how to resolve such conflicting interpretations. Contact the NFPA whenever a conflict that cannot be resolved occurs.

Most local inspection authorities have no jurisdiction over building construction on projects owned or leased by the Federal government. Each of the various

Federal agencies establishes its own inspection procedures and requirements covering both the type and quality of material and its installation, every agency also has its own design and inspection departments. Although all governmental agencies usually base their requirements on the NEC, they may differ slightly from the NEC's minimum requirements.

All electrical workers, especially those in a supervisory capacity, should familiarize themselves with the inspection and installation regulations of the particular government agency having jurisdiction over any building or project with which they may be concerned from an installation or management standpoint.

1-7   TOOLS OF THE TRADE

High-quality work in any trade or profession can be accomplished only by the correct use of high-quality tools; the electrical industry is no exception. A knowledge of and the ability to use hand tools properly is the electrician's stock-in-trade since nearly every operation performed by the electrical worker requires the use of some hand tool.

The hand tools commonly used by the electrician as set forth in the majority of labor agreements are listed in Table 1–1. This table lists the minimum essential tools needed in order to perform good work on the usual kind of electrical installations. The tools should be of the best quality available. An electrician should not only know how to use them properly, but also how to care for them.

Table 1–1   TOOLS FOR INSIDE ELECTRICAL WORKERS

1 Tool box	1 1/2″ Wood chisel
1 8″ Side cutters (Figure 1–14)	1 6″ Crescent wrench (Figure 1–18)
2 Pr. channel locks	1 10″ Crescent wrench (Figure 1–17)
1 10″ Screwdriver (Figure 1–19)	1 50′ Steel tape
1 6″ Screwdriver	1 Key hole saw
1 Claw hammer (Figure 1–20)	1 Phillip's screwdriver
1 6′ Folding rule (Figure 1–22)	1 Gripping screwdriver
1 Inexpensive voltage tester	1 10″ Tin snips
1 2″ Conduit reamer	1 Small architect's scale
1 Brace	1 Electrician's knife
1 Expansion bit	1 8″ Level (Figure 1–21)
1 Bit extension	1 Diagonal pliers (Figure 1–15)
1 Combination square	1 Long nose pliers (Figure 1–16)
1 Center punch	1 Lock
1 ½″ Chisel	1 Hacksaw frame (Figure 1–23)
1 Flashlight (preferably wired for testing continuity)	1 10″ Mill file
	1 Fuse puller
1 Tap wrench	

QUESTIONS

Check the correct answer for the following True—False Questions. If any part of the statement is false, the statement should be marked "False."

T   **F**

1   The working drawings and specifications prepared by an architect/engineer for a new building are used for bidding purposes only.   _____ _____

2   Fundamental electrical design procedures is important to all electrical workers, and not only to the design engineer.   _____ _____

3   The general contractor normally hires subcontractors to handle special work, such as plumbing, heating and cooling, and the electrical work.   _____ _____

4   Subcontractors—like electrical contractors—are responsible for furnishing all labor and materials for completing their portion of the building contract but are *not* responsible for the performance of their phase of the work.   _____ _____

5   In the construction of a new building, an architectural firm is responsible only for the preparation of the working drawings; the engineer always prepares the written specifications.   _____ _____

6   Shown below are 20 symbols commonly found on electrical working drawings. In the space provided, place the number corresponding to the correct answer found in the list.

1	_____	A.	Telephone outlet, wall-mounted
2	_____	B.	Smoke detector
3	_____	C.	Fire-alarm striking station
4	_____	D.	Fan coil-unit connection
5	_____	E.	Telephone conduit
6	_____	F.	Home run to panel
7	_____	G.	Conduit, exposed
8	_____	H.	Disconnect safety switch
9	_____	I.	Light or power panel
10	_____	J.	Switch, three-way
11	_____	K.	Receptacle, floor type
12	_____	L.	Combination switch and receptacle
13	_____	M.	Receptacle, weatherproof
14	_____	N.	Receptacle, duplex–grounded
15	_____	O.	Indicates lighting fixture type
16	_____	P.	Exit light, wall-mounted
17	_____	Q.	Exit light, surface-mounted

18   _____   R.   Incandescent fixture, wall-mounted

19   _____   S.   Incandescent fixture, recessed

20   _____   T.   Florescent fixture

The following questions should be answered by filling in the blanks.

7   Drawings that represent the physical arrangement and views of specific electrical equipment are called _____ _____ drawings.

8   A complete set of electrical working drawings usually consists of:

     a. _____ plan
     b. _____ plan
     c. _____ diagrams
     d. _____ diagrams

9   In general, written electrical specifications show the _____ of materials to be used on the project and the _____ in which the electrical system shall be installed.

10   The National Electrical Code is published by the _____ _____ _____ _____.

11   The NEC is a result of the best efforts of _____ _____, _____ _____, _____, and other concerned experts throughout the country.

12   The American Insurance Association has established a testing organization for electrical equipment known as _____ _____; the abbreviation on their label is _____.

13   In the space provided, place the proper name for each tool shown in the figures below.

     a. Figure 1–14   _____   f. Figure 1–19   _____
     b. Figure 1–15   _____   g. Figure 1–20   _____
     c. Figure 1–16   _____   h. Figure 1–21   _____
     d. Figure 1–17   _____   i. Figure 1–22   _____
     e. Figure 1–18   _____   j. Figure 1–23   _____

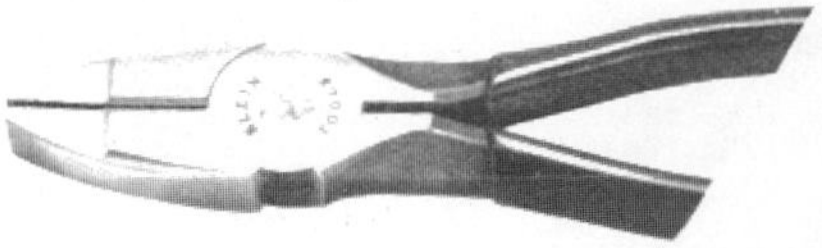

Figure 1–14 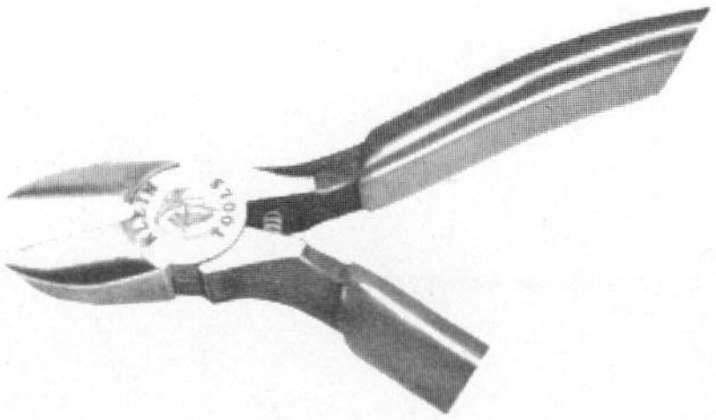   Figure 1–15

Figure 1–16

Figure 1–17

Figure 1–18

Figure 1–19

Figure 1–20

Figure 1–21

Figure 1–22

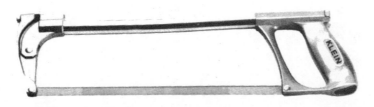

Figure 1–23

2

HOW TO USE
THE NEC

In order to understand the language and terms used in the NEC as well as in the industry in general, those involved in the electrical construction industry should carefully read the definitions listed in Chapter 1, Article 100 of the Code. For simplicity, this article lists only definitions essential to the proper use of the Code; only terms used in two or more other articles are defined in full in this article. Other definitions, however, are defined in the individual articles of the NEC in which they apply.

2–1   GENERAL CONSIDERATIONS

Mandatory rules of the NEC are characterized by the use of the word **shall**, while *advisory* rules are related by the use of the word **should**. When statements using the latter word are made, they are stated as *recommendations* of what is advised but *not necessarily required*. Some localities, however change the wording from ''should'' to ''shall'' in their local ordinances, meaning that the ordinance *must* be complied with. When working in a new area, it is therefore useful to find out if there are additional legal requirements amending the NEC and, specifically, what these are.

General requirements for electrical installations are given in Article 110. This is a short article, and it can be fully read in less than five minutes. This article, along with Article 100, should be read over several times until the information contained in both is fully understood and firmly implanted in the reader's mind. With a good understanding of this basic knowledge, the reader will find the remaining portions of the NEC easier to understand.

2–2   WIRING DESIGN AND PROTECTION

Chapter 2 of the NEC, ''Wiring Design and Protection,'' is the chapter that most electrical workers use most often. It covers data such as use and identification of grounded conductors, branch circuits, feeders, calculations, services, overcurrent protection, and grounding: all of which are necessary for *any* type of electrical system regardless of the type of building in which the system is installed.

Since most electrical designers and workers refer to Chapter 2 of the NEC so often, it does not take long for most of them to practically memorize it. However, the beginner constantly will have to look up various items before they become committed to memory.

Note that Chapter 2 deals with *how-to* items; that is, how to provide proper spacing for conductor supports, how to provide temporary wiring, how to size the proper grounding conductor or electrode, etc. If a problem develops pertaining to the design or installation of a conventional electrical system, you can usually find the answer in Chapter 2 of the NEC. Thumb through the pages and glance at the article headings until you find a heading under which your particular question or problem should appear, then read through the entire article. If you cannot find the answer there, refer to the index to find other articles in the NEC that cover your particular problem.

2–3   WIRING METHODS AND MATERIALS

The wiring method used on a given electrical installation is determined by the type of building construction, the location of the building and the location of the wiring in it, the type of atmosphere in the building or the area surrounding the building, mechanical factors, the relative costs of different wiring methods, etc. Rules governing this phase of the electrical installation are found in Chapter 3 of the NEC, ''Wiring Methods and Materials.'' The provisions of this article apply to all wiring installations except remote-control switching (Article 725), low-energy power circuits (Article 725), signal systems (Article 725), communication systems, and conductors (Article 800), which form an integral part of equipment such as motors and motor controllers.

In general, there are three basic wiring methods used in the majority of modern electrical systems. They are:

1   Sheathed cables of two or more conductors, such as NM cable and BX armored cable (Articles 330 to 339)

2   Raceway wiring systems like rigid and EMT conduit (Articles 342 to 358)

3   Busways (Article 364)

Nearly all wiring methods are a variation of these three basic methods.

Because there is a wide range of sizes and forms of electrical conductors, electricians should carefully note whether or not a particular type is specified on the drawings or specifications. If none is specified, the electrician should choose the tyye and size under NEC requirements that provides the most economical installation for the purpose to be served. A complete description of all types of electrical conductors may be found in Chapter 3 of the NEC (Articles 310–313).

Rules pertaining to raceways, boxes, cabinets, and raceway fittings are also found in this section of the NEC (Articles 318 to 384). Outlet boxes vary in size and shape, depending on their use, the size of the raceway, the number of conductors entering the box, the type of building construction, atmospheric condition of the area, and other special requirements. Chapter 3 should answer most questions pertaining to the selection and use of these items.

It is not within the scope of the NEC (nor this book) to describe in detail all the types and sizes of outlet boxes. However, manufacturers of outlet boxes have excellent catalogs illustrating all of their products with which electricians must be familiar. It is recommended that the reader obtain catalogs of outlet boxes as well as of other types of electrical materials.

In order to control electrical circuits or to connect portable equipment to electrical circuits, a wide variety of switches, push buttons, pilot lamp receptacles, convenience outlets, etc., are available. Chapter 3 (Article 380) covers these items in general. To become familiar with all the various types available, the reader should obtain catalogs from the manufacturers of such wiring devices.

Switchboards and panelboards are also covered in Chapter 3 (Article 384). Items such as location, installation methods, clearances, grounding, and overcurrent protection are also covered in this section.

2–4   EQUIPMENT FOR GENERAL USE

Chapter 4 of the NEC begins with the use and installation of flexible cords and cables, including the trade name, type letter, wire size, number of conductors, conductor insulation, outer covering, and use of each. The chapter continues on to fixture wires—again giving trade name, type letter, and other pertinent details.

Article 410 on lighting fixtures is of especial interest to electricians because it gives the installation procedures for the various types of fixtures for use in specific locations (fixtures near combustible materials, in closets, etc.). It is not within the scope of the NEC nor this chapter to go into the details of lighting design, but other chapters in this book will provide the reader with a basic knowledge of illumination and lighting calculations. The manufacturers of lighting equipment provide a great amount of illuminating data that can be used to advantage by the electrician.

The section on electric motors is Articles 430 through 445 of the NEC, including mounting the motor and making electrical connections to it. Air-conditioning and heating equipment, transformers, and capacitors are covered in Articles 440 through 460.

Although storage batteries are not often thought of as part of an electrical wiring system for building construction, they are often used to provide standby emergency lighting service and to supply power to security systems separate from the main a.c. electrical system. Most requirements pertaining to battery-operated systems will be found in Chapter 4 (Article 480).

2–5   SPECIAL OCCUPANCIES

Any area where the atmosphere or any material in the area is such that the sparking of operating electrical equipment may cause an explosion or fire is considered a hazardous location. These areas are covered in Chapter 5 of the NEC, "**Special Occupancies**." Typical locations include commercial garages, aircraft hangers, and service stations. These locations have been classified in the NEC into three class locations. Various atmospheric groups have been established on the basis of the explosive character of the atmosphere for the purpose of testing and approving equipment for use in the various groups (Articles 500–501).

The basic principle of explosion-proof wiring is to design and install such a system that, when the inevitable arcing occurs within the electrical system, the ignition of the surrounding explosive atmosphere is prevented. The basic principles of such an installation are covered in Chapter 5 of the NEC (Articles 501–4, 502–4, and 503–3).

Areas containing flammable gases or vapors in the air fall under Class I locations (Article 501). Such locations include paint spray booths, dyeing plants where hazardous liquids are used, and gas generator rooms.

Class II locations (Article 502) include areas where combustible dust is present: grain handling and storage plants, dust and stock collector areas, sugar-pulverizing plants, and other similar locations where combustible dust may, under normal operating conditions, be present in the air in quantities sufficient to produce explosive or ignitible mixtures.

Class III locations (Article 503) are those that are hazardous because of the presence of easily ignitible fibers or flyings but in which such fibers or flyings are not likely to be in suspension in air in quantities sufficient to produce ignitible mixtures. Typical locations are cotton mills, rayon mills, and clothing manufacturing plants.

Garages and similar locations where volatile or flammable liquids are used are not usually considered critically hazardous locations. However, sufficient hazard may exist to require certain precautionary measures to be taken in the electrical installation. The NEC (Articles 511 and 514) bases it regulation for these areas on the theory that volatile gases will be confined to an area not more than four feet above the floor. Therefore, in most cases, conventional raceway systems are permitted above this level. When conditions indicate that the area concerned is more hazardous than usual, the applicable type of explosion-proof wiring (including seal offs) may be recommended or required.

Theaters and similar occupancies also fall under the regulations set forth in Chapter 5 of the NEC (Article 520). Because hazards to life and property due to fire and panic exist in theaters, there are certain requirements in addition to those for usual commercial wiring installations. While drive-in theaters do not present the inherent hazards of enclosed auditoriums, the projection rooms and areas adjacent to these rooms must be properly ventilated and wired for the protection of operating personnel and others using the areas.

Other areas falling under the regulation of Chapter 5 of the NEC include: residential storage garages, aircraft hangars, service stations, bulk-storage plants, finishing processes, health care facilities, mobile homes and parks, and recreation vehicles and parks.

2–6   SPECIAL EQUIPMENT

The provisions in Chapter 6 of the NEC apply to electric signs and outline lighting, cranes and hoists, elevators, electric welders, and sound-recording and similar equipment. Therefore, any electrician who works on any of these special systems should thoroughly check Chapter 6 along with other applicable chapters.

Electric signs and outline lighting (Article 600) are usually considered to be self-contained equipment installed outside and apart from the building wiring system. However, the circuits feeding these lights are usually supplied from within the building itself. Therefore, some means of disconnecting such equipment from the supply circuit is required. All such equipment must be grounded except when it is insulated from the ground or conducting surfaces or is inaccessible to unauthorized persons.

Neon tubing (Article 600–633), where used, requires the use of step-up transformers to provide the necessary operating voltages, and secondary conductors must, of course, have an insulation and be rated for this high voltage; terminators must be of the proper type and be protected from or inaccessible to unqualified persons. In most instances, the electrician will be responsible only for providing the feeder circuit to the location of the lights; qualified sign installers will usually do the actual sign work.

Cranes and hoists (Article 610) are usually furnished and installed by trades other than electricians. However, it is usually the electrician's responsibility to furnish all wiring, feeders, and connections for the equipment. Such wiring will consist of the control and operating circuit wiring on the equipment itself and the conductors that supply electric current to the equipment in a manner allowing it to move or operate properly. Furthermore, electricians are normally required to furnish and install the contact conductor and necessary suspension and supporting insulators. Motors, motor control equipment, and similar items are normally furnished by the crane manufacturer.

The majority of the electrical work involved in the installation and operation of elevators, dumbwaiters, escalators, and moving walks (Article 620) is usually furnished and installed by the manufacturer. The electrician is usually required only to furnish a feeder terminating in a disconnect means in the bottom of the elevator shaft and perhaps to provide a lighting circuit to a junction box midway in the elevator shaft for connection of the elevator cage lighting cable. Articles in Chapter 6 of the NEC give most of the requirements for these installations.

Electric welding equipment (Article 630) is normally treated as a piece of industrial power equipment for which a special power outlet is provided. There are certain specific conditions, however, that apply to circuits supplying welding equipment and that are outlined in Chapter 6 of the NEC.

Wiring for sound-recording and similar equipment (Article 640) is essentially of the low-voltage type. Special outlet boxes or cabinets are usually provided with the equipment although some items may be mounted in or on standard outlet boxes. Some systems of this type require direct current, which is obtained from rectifying equipment, batteries, or motor generators. The low-voltage alternating current is obtained through the use of relatively small transformers connected on the primary side to a 120-volt circuit within the building.

Besides becoming familiar with the provisions set forth in this chapter of the NEC, the electrician should obtain (in advance) rough-in drawings of the equipment to be installed. This can save many hours of grief when it is found after the building finishes have been completed that the wrong outlet box was installed or that it was improperly placed.

Other items covered in Chapter 6 of the NEC include X-ray equipment (Article 660), induction and dielectric heat-generating equipment (Article 665), and machine tools (Article 670). A brief reading of this chapter will provide a sound basis for approaching such work in a professional manner, and then problems can be studied in more detail during the installation.

2–7   SPECIAL CONDITIONS

In most commercial buildings, codes and ordinances require a means of lighting public rooms, halls, stairways and entrances so that there will be sufficient light to allow the occupants to exit from the building if the general building lighting is interrupted. Exit doors must be clearly indicated by illuminated exit signs.

Articles in Chapter 7 of the NEC give provisions for installing emergency lighting systems. Such circuits should be arranged in such a manner that they may be automatically transferred to an alternate source of current supply—from storage batteries, gasoline-driven generators, or properly connected to the supply side of the main service—so that disconnecting the main service switch will not disconnect the emergency circuits. Additional details may be found in Article 700 of the NEC.

Circuits and equipment operating at more than 600 volts between conductors will also be found in this portion of the Code (Article 710). In general, conductor insulations must be of a type approved for the operating voltage, and the conductors must be installed in rigid conduit, duct or armored cable approved for the voltage used. These cables must be terminated with approved cable terminating devices. All exposed live parts must be given careful attention and be adequately guarded by suitable enclosures or isolated by elevating the equipment beyond the reach of unauthorized personnel. Overcurrent protection and disconnecting means must be manufactured specifically for the operating voltage of the system. Examples of such high-voltage installations include feeders for synchronous motors and capacitors, substations, and transformer vaults. Other details are covered in articles in Chapter 7 of the NEC.

Among other items covered in Chapter 7 of the NEC is the installation of outside wiring other than for electric signs. Such wiring is attached to the building, between two or more buildings, run overhead, underground, or in a raceway fastened to the face of the building (Article 730).

Overhead systems may consist of individual conductors supported on or by insulators on or at the building surface, or by means of approved cable either attached to or suspended between buildings. When the buildings are some distance from each other, intermediate supporting poles are necessary, and certain clearance distances must be maintained over driveways and building structures.

Underground wiring between buildings or outside of buildings is installed either directly in the ground as direct-burial cable or else is pulled through raceways consisting of rigid conduit, PVC conduit, or ducts encased in concrete.

2–8   COMMUNICATION SYSTEMS

Chapter 8 of the NEC deals with communication systems and circuits, i.e., telephone, telegraph, district messenger, fire and burglar alarms, and similar central station systems as well as telephone systems not connected to a central station system but using similar types of equipment, methods of installation, and maintenance. Articles in this chapter also cover radio and television equipment, community antenna television and radio distribution systems (cable TV), and similar systems. The basic requirements are outlined in this chapter of the Code, but a good knowledge of communication systems is also required for a proper installation.

2–9   USING THE NEC

The drawing in Figure 2–1 illustrates the NEC requirements for a complete electrical distribution system from the generating plant through conversion and transmission to distribution and utilization. The use of this illustration is briefly explained below.

Assume that an electrician is required to wire a residential building or dwelling. In most occupancies of this type, the local power company will furnish electric service to a point on the building (from the service drop to the meter socket). The owner is then required to furnish a service entrance (cable or conduit) to the meter base and from the meter to a disconnect switch or panelboard inside the building. The owner is responsible for all interior wiring, such as power and lighting outlets, branch circuit wiring, and feeder (see Figure 2–1).

Usually the electrician will first lay out the various outlets and then size the branch circuits and service entrance. Note in the diagram (in the lower right-hand corner under ''Domestic and General Power'') that Articles 210, 220, and 300 are given as reference at the service entrance conduit.

Article 210—Branch Circuits—covers the general layout and requirements of all branch circuits within the house, such as the spacing of duplex receptacles in residential occupancies. Knowledge of these requirements is necessary in order to calculate properly the size of the service entrance. Article 220 gives the procedure for calculating the size of the service entrance as well as branch circuits and feeders. Article 300 informs the electrician about the correct wire size and the type of insulation required.

Inside the house, Articles 230 and 250 of the NEC (as shown in the diagram) outline the requirements for circuit breakers or fusible service equipment and load centers. Again, Articles 210 and 220 are used for reference about sizes and installing branch circuits, Article 410 for lighting fixtures, and Article 422 for appliances. Article 800 deals with TV antennas and outlets as well as other communications system.

Under ''Low-Voltage Industrial and Commercial Power'' in the illustration, additional articles are given for switchboards, busways, capacitors, transformers, etc. Until the NEC becomes second nature, the electrician should keep this diagram handy for reference when working on any type of building or electrical system. Although the diagram is not complete, it does form a good basis for quickly locating certain articles in the NEC.

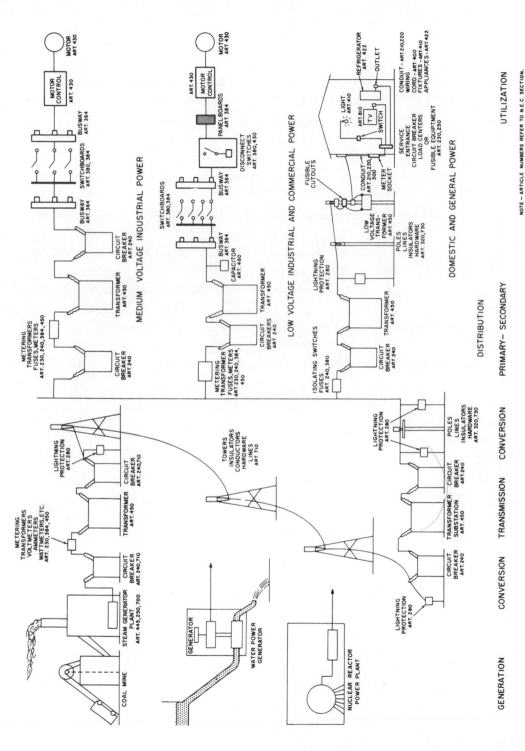

Figure 2-1 COMPLETE ELECTRICAL DISTRIBUTION SYSTEM FROM THE GENERATING PLANT THROUGH CONVERSION AND TRANSMISSION TO DISTRIBUTION AND UTILIZATION.

QUESTIONS

The following questions should be answered by filling in the blanks.

1   Rules governing types of wiring methods to be used in a given electrical installation are mostly found in Chapter _____ of the NEC.

2   Mandatory rules of the NEC are characterized by the use of the word _____ whereas advisory rules or recommendations are characterized by the use of the word _____.

3   If an electrician wants to locate the requirements for providing temporary wiring, he should first refer to Chapter _____ of the NEC.

4   A complete description of all types of electrical conductors may be found in Chapter _____ of the NEC.

5   The NEC does not describe in detail all the types and sizes of outlet boxes, but a complete description may be obtained from the _____.

6   Emergency lighting systems normally obtain their auxiliary power from _____ or _____.

7   Any area where the operation of electrical equipment may cause an explosion or fire is considered to be a _____ location.

8   Provisions for the wiring of electrical signs is outlined in Chapter _____ of the NEC.

9   Areas containing flammable gases or vapors in the air fall under Class _____ locations.

10   If an electrician is selecting and installing a circuit breaker enclosure in an industrial electrical system, he would first refer to Article _____ of the NEC, where the provisions are outlined.

3

PLANNING AN ELECTRICAL INSTALLATION

Very few electrical installations are exact duplicates of any other installation because most are specifically designed to fit the building structure or the owner's requirements. However, most electrical installations for building construction follow a set pattern, using certain general steps to be followed during the installation, regardless of the type of building or wiring method used.

3–1   FACTORS TO CONSIDER

In planning an electrical installation, certain factors should be taken into consideration. These include:

1   Wiring method

2   Overhead or underground electrical service

3   Type of building construction and/or occupancy

4   Type of service entrance and equipment

5   Voltages to be used

6   Grade of wiring devices, lighting fixtures, etc.

7   Selection of lighting fixtures

8   Type of heating and cooling system

9   Control wiring for heating and cooling system

10   Signal and alarm systems

On larger types of building construction, complete working drawings and specifications are usually provided for the electricians. On projects for which complete plans and specifications have not been prepared, experienced electricians still readily recognize, within certain limits, the type of system that is required. However, the foreman in charge of electrical construction should always check the local ordinances when selecting a wiring method. Even if plans and specifications have been provided by an architect or engineer, local regulations should be checked. If more than one wiring method may be practical, a decision must be made prior to the installation about which type to use.

In a residential occupancy, for example, electricians know that a 120/240-volt, single-phase electric service is invariably provided by the local utility company. They also know that the service and feeder wiring is usually three wire, that all branch circuits are either two or three wire, and that the safety switches, service equipment, and panelboards are three-wire solid neutral. On each project, however, the electrical contractor or his electricians must determine where the point of service-drop attachment to the building is to be located and whether the service feeder is to be installed as part of the electrical contract or by the utility company.

A study of the building, its construction documents, or both should be made before the wiring methods and materials are selected. Such surveys are generally necessary and are made by the electrician on small jobs; the architect or engineer usually takes care of this responsibility on larger projects. On some projects, the electrical drawings may be very vague—merely showing the location of the outlets and service-entrance; the wiring details are left to the electrician on the job who has to "design" the installation to comply with the NEC and to determine the circuit arrangement to be followed.

The information obtained from a study of the working documents or survey should include the materials of which the building is constructed, i.e., frame, masonry, etc. It should further indicate interior spaces available for raceways and cable systems, the mechanical equipment of other trades that will have to be avoided, the room finishes, and similar items.

Electricians working on the project should not only examine the electrical drawings, but also the architectural floor plans, sections, and details in order to obtain a better overall picture of the building construction.

3–2   WIRING METHODS

Every electrician should have the basic knowledge of conventional wiring methods (as described in Chapter 4 of this book) before attempting to plan an electrical installation. However, the following is a summary of the different wiring methods normally used in various types of occupancies (from residential to hazardous locations).

Residential

Service entrance—Service-entrance cable is used in the majority of the cases. Rigid conduit is used for service mast extending through roofs and for underground services.

Feeders—Metal-clad (BX) or nonmetallic (Romex) cables are the most popular wiring methods. Type SE cable is sometimes used for the electric range, dryer, and water heater circuits.

Branch circuits—Romex and BX cables are used almost exclusively. Type UF cable may be used for direct burial in the earth for feeding outdoor outlets. EMT is also used (infrequently) for exposed wiring in basements, garages, etc.

Small Commercial Occupancy

Service entrance—Rigid metallic conduit is used for overhead service, whereas rigid metallic or PVC conduit is used for underground service.

Feeders—Rigid conduit and EMT are employed. Flexible metallic conduit may be required for connections to certain electrical motor-driven equipment.

Branch circuits—Type AC (BX) cable is used for wiring in nonmasonry partitions and for lighting fixtures. Rigid conduit or EMT is used for in-slab circuits and wiring on or in masonry walls.

Large Commercial Occupancy

Service entrance—Rigid metallic conduit is used for overhead service drops, and rigid metallic or PVC conduit is employed for underground service laterals.

Feeders—Rigid conduit or EMT is used; electrical drawings and specifications should be checked.

Branch circuits—Type AC cable and EMT are used for wiring in ceiling; rigid conduit is used for slab work and exposed raceways.

Industrial Plant

Electric Service—Rigid metallic conduit is used for overhead services while fiber ducts, encased in concrete are used for underground service; a substation may also be involved.

Feeders—Rigid conduit, busways, cable trays, and MI cable are used for most feeders.

Branch circuits—Rigid conduit, EMT, and cable trays are the wiring method most often used.

Control Wiring—Rigid conduit, EMT, multiconductor cable assemblies, and cable trays are used for control wiring.

Outdoor Sports Lighting

Electric Service—Combination of rigid metallic conduit and direct-burial installations are quite common.

Feeders—Rigid conduit and direct burial UF or USE cable are used.

Hazardous Locations

All wiring in hazardous locations is done exclusively in rigid conduit using explosion-proof or dust-proof fittings depending on the type of hazardous location. See Chapter 12.

3–3   TYPE AND SIZE OF ELECTRIC SERVICE

Service conductors should be rated sufficiently to carry the complete, computed electrical load of the building and any anticipated future loads. The electrician should work closely with the local utility company to determine the optimum service-entrance location as well as the proper location for metering the energy consumed.

Chapter 11 details calculations of electric services for several types of occupancies. In general, the required service is computed on definitely located outlets and loads for which the full load in watts or amperes can be assigned for wiring design purposes. However, in some cases, the individual load cannot be determined precisely until the space or building is occupied. In such cases, the anticipated load must be computed on a watts per quare foot basis, using tables based on experience with similar buildings, as per NEC (Article 220).

After the loads have been computed, panelboard requirements can be determined (Article 384). The number and sizes of feeders needed to supply these panels can then be specified.

The NEC allows certain demand factors to be applied to the total computed load in the calculation of feeder and service requirements. The *demand factor* is the ratio of the maximum demand of a system to the total connected load of the system under consideration. According to the NEC, such a demand factor may be applied in certain occupancies if the total connected load exceeds a certain value.

Example 3–1

A warehouse is determined to have a total connected-lighting load of 83,200 watts. Find:

a   The demand factors that may be applied to this load.

b   The portion of load computed at the demand factor.

c   The total demand to be used for sizing lighting feeders.

Solution

a   Table 220–4(b) of the NEC states that for warehouses the first 12,500 watts must be computed at 100 percent and the remaining load at a demand factor of 50 percent.

b   Therefore, 83,200 watts − 12,500 watts = 70,700 W.

c   The total demand is then $(0.5 \times 70,700 + 12,500)$ W = 47,850.

Other examples will be given in Chapter 11.

3–4   SELECTION OF LIGHTING FIXTURES

On most larger buildings, the lighting requirements are calculated by an illumination engineer, and the number and types of lighting fixtures are specified in the working documents. On smaller commercial and most residential occupancies, the electrician will be required to determine an adequate lighting plan.

The basic requirement for any lighting design is to provide the highest level of visual comfort and performance that is consistent with the type of area to be illuminated and with the budget provided by the owners. Details of lighting design and application are given in Chapter 7 of this book.

Once the number and types of lighting fixtures have been selected, other items, such as mounting attachments and wiring methods, must be determined.

3–5   TYPES OF HEATING AND COOLING SYSTEMS

The type of heating and cooling system used in a particular building has a great bearing on the electric service and feeders. For example, if electric heating is used in place of, say, an oil burner, the electric load could increase many times. Therefore, it is very important for the electrician to cover the details of the heating and cooling system as well as of any other equipment requiring electric power.

The mechanical contractor will normally handle the installation of all air-conditioning and heating equipment, but the electrician must know the electrical specifications in order to plan the wiring system, feeders, etc. The electrician must also know certain details on the control wiring if he is to connect this wiring.

3–6   SIGNAL AND ALARM SYSTEMS

The desired locations of all telephone, intercom, alarm, and other signal or communication outlets must be selected before beginning an electrical installation. If the electrical contractor is dealing directly with an architect or owner, he should be prepared to discuss the benefits of such systems along with approximate costs.

The type and size of the control wiring will have to be determined along with miscellaneous items in order to complete the system—especially during the rough-in stage. Chapter 14 covers signal and communication wiring.

3-7   REQUIRED CALCULATIONS

Whether a building project is designed by an electrical engineer or by the electrician himself, there are times when calculations will have to be made during electrical installation. A customer may ask the electrician to connect an electric range rated at 12,000 watts. What wire size should be used? The electrician will have to divide the total wattage by the line-to-line voltage and then multiply by the demand factor given in Table 220–5 of the NEC.

Perhaps the owner of a small store is having an old coal-fired furnace replaced by an electric heat pump. This will require a larger electric service. The electrician must know how to calculate the additional load required and determine the size of the service.

Chapter 5 of this book covers most calculations that are needed on conventional electrical systems.

3-8   LAYOUT OF THE SYSTEM

1   Mark the location of all the switch, lighting, convenience, and minor power outlets. Measure and mark the correct mounting height above the finished floor.

2   Determine the best route for the branch circuit wiring and how wiring might be saved by utilizing a common neutral for up to two circuits on single phase and as many as three circuits on three-phase systems.

3   Estimate the materials needed for the branch circuit runs in case they are to be purchased or picked up from the shop.

4   Locate and account for all outside wiring to items such as outside floodlights and signs.

5   Determine how the service equipment will be arranged, and make certain that all branch circuits and feeders can be routed to the panels and subpanels with the least amount of conduit and wire.

6   Locate and account for all large power equipment, such as air conditioners, compressors, and motors.

7   Contact the local power company for any possible problems with your layout in regard to their bringing in electric service.

8   Make a complete list of materials from the information obtained.

9   Estimate the type of tools required for the installation, especially if the installation is unusual. Examples include installing MI cable and wiring in hazardous locations.

3–9   SUMMARY

The wiring system of any building must be designed to meet certain immediate needs but should also be sized to serve future anticipated load.

The electrician should become familiar with the requirements either by performing a survey of the building (in the case of an existing structure) or by carefully studying the construction documents (plans and specifications). Surrounding conditions and the location should also be observed in order to simplify the installation of electrical service. The architect's or owner's conception of any special needs, the atmospheres in which the wiring will be used, local ordinances, etc., are among the things to consider in arriving at a general understanding of the projected installation.

QUESTIONS

Answer the following questions by filling in the blanks.

1   One type of occupancy that probably will not have complete working drawings and specifications for the electrician to follow is _____ buildings.

2   The term used to indicate diversity in an electrical load is _____ factor.

3   In a residential occupancy, the electric service will normally be a _____-volt, single-phase service.

4   The information obtained from the working drawings or building survey should include the _____ of which the building is constructed.

5   The wiring method normally used for an overhead electric service for a large commercial building is _____.

6   All wiring in hazardous locations will utilize rigid conduit and _____ or _____ fittings.

7   Article _____ of the NEC gives details on sizing the electric service.

8   The demand factor is the _____ of the maximum demand of an electrical system to the total connected _____.

9   The _____ _____ contractor will normally install large air-conditioning and heating equipment.

Circle the proper answer.

10   Electrical calculations are required:

    a   only if construction documents are not available from the engineer.

    b   only on residential occupancies.

    c   on almost all electrical projects.

    d   in designing electrical systems only.

PROBLEMS

1   A hospital has a total lighting load of 100,000 watts. Calculate the total lighting demand for sizing the lighting feeder.

2   A multistory apartment house has a total lighting load (exclusive of electric cooking and appliances) of 150,000 watts. Calculate the total lighting demand for sizing the lighting feeder.

4

WIRING METHODS

Conductors of copper or aluminum[1] are used almost exclusively for interior wiring, although steel bus-bars coated with zinc—copper plating are sometimes used to meet special conditions. Copper or aluminum wires may be either stranded (Figure 4–1a) or solid (Figure 4–1b). In most cases, conductors sized No. 10 AWG and larger are stranded. However, there are exceptions; for example, fixture wires and flexible cords (No. 14 AWG and smaller) are normally stranded, and conductors used for outside overhead wiring will sometimes be solid for sizes No. 8 AWG and smaller.

Conductors, whether stranded or solid, should not be smaller than No. 14 AWG, except for certain limited applications: for printing-press control circuits (Section 400–7); for fixture wire (Section 410–8); for fractional horsepower motors (Section 430–22); for cranes and hoists (Section 610–14); for elevator control and signal circuits (Section 620–12); and for remote-control, low-energy power, low-voltage power and signal circuits (Section 725–13).

Figure 4–1a   CROSS SECTION OF
STRANDED ELECTRICAL WIRE.

Figure 4–1b   CROSS SECTION OF
SOLID ELECTRICAL WIRE.

[1]Use of aluminum conductors for interior residential branch-circuit wiring is not recommended for installations because of fire hazard.

Because most round wires are relatively small in diameter and cross-sectional area, the diameter is usually measured in mils and is represented as D. One mil = 0.001 inches, and therefore D in mils = 1000 $\times$ D in inches. The cross-sectional area of a conductor is proportional to the square of the diameter of the conductor:

$$A = \frac{\pi}{4} D^2, \text{ so } A \propto D^2$$

From this, we define the area, in circular mils, as

$$A = D^2 \text{ cm} \tag{4-1}$$

where D is the diameter in mils, and A is the cross-sectional area in circular mils.

Example 4–1

A solid wire has a diameter of 0.5 inches. Calculate:

a The diameter in mils.

b The circular mil area of the wire.

Solution

a $D_{mils} = 1000 \times D_{inches}$
$= 1000 \times 0.5 = 500$ mils

b $A = D^2 = (500 \text{ mils})^2 = 250{,}000 \text{ CM} = 250 \text{ MCM}$

Wires are sized both by CM area and either AWG number or area in MCM as shown in Table 8, Chapter 9 of the NEC. An examination of this table shows:

1 As the AWG number *decreases*, the diameter *increases*.

2 AWG numbers are used from No. 18 (smallest) down to No. 4/0 (0000), which is 211,600 CM or approximately 212 MCM.

3 Beyond No. 4/0, the wire size is shown *directly* in MCM.

4 The largest size conductor is 2000 MCM, having a diameter of 1.63 inches.

Example 4–2

A wire has an area of 105,600 CM. Calculate:

a The wire diameter in mils.

b The wire diameter in inches, assuming the conductor is solid.

c   Compare your values with those in Table 8, Chapter 9 of the NEC, and account for differences.

Solution

a $D = \sqrt{A} = \sqrt{105{,}600} \text{ CM}$
      $= 324.96 \text{ mils} = 325 \text{ mils}$

b $D_{inches} = D_{mils}/1000 = 325/1000 = 0.325$ inches, if the conductor is solid.

c D_{inches} from Table 8, Chapter 9 of the NEC is 0.372 inches. Since the commercial conductor consists of 19 stranded wires, the diameter of the commercial conductor is effectively larger than a solid conductor of the same CM area.

The larger the CM area of a conductor, the more current the conductor can carry without damage to the conductor or its insulation. The current-carrying capacity (ampacity) is also affected by the type of insulation, the operating temperature, and whether it is used in free air or is enclosed in a raceway. Obviously, the more the conductor is insulated and enclosed in a raceway, the higher the temperature that limits its current-carrying capabilities.

Table 310-2 of the NEC shows the various conductor insulations and their maximum operating temperature along with their common applications. Tables 310-12 and 310-13 show the allowable ampacities of copper conductors in a raceway (not more than 3) and in free air, respectively. Table 310–12 (Figure 4–1c) also applies to conductors buried in the earth. Note that the ampacity in free air is always greater than the ampacity of conductors in a raceway or buried in the earth. This is due to the cooler operating temperature of conductors in free air.

If more than three conductors are used in a raceway or buried in the earth, their ampacity must be derated according to Table 4–1. A neutral conductor that carries only the unbalanced load from other conductors need not be counted. For example, a 120/208-volt, 4-wire feeder enclosed in a raceway need not be derated because there are only *three* phase conductors; the fourth wire is the neutral used to carry the unbalanced load.

Table 4–1   REDUCTION IN AMPACITY FOR 4 OR MORE CONDUCTORS IN A RACEWAY

Number of conductors	Value in tables 310–12 and 310–14
4–6	80%
7–24	70
25–42	60
43 and above	50

AWG MCM	60°C (140°F) TYPES RUW (14-2), T, TW	75°C (167°F) TYPES RH, RHW, RUH (14-2), THW, THWN, XHHW	85°C (185°F) TYPES V, MI	90°C (194°F) TYPES TA, TBS, SA, AVB, SIS, FEP, FEPB, RHH, THHN, XHHW**	110°C (230°F) TYPES AVA, AVL	125°C (257°F) TYPES AI (14-8), AIA	200°C (392°F) TYPES A (14-8), AA, FEP*, FEPB*	250°C (482°F) TYPE TFE (Nickel or nickel-coated copper only)
14	15	15	25	25†	30	30	30	40
12	20	20	30	30†	35	40	40	55
10	30	30	40	40†	45	50	55	75
8	40	45	50	50	60	65	70	95
6	55	65	70	70	80	85	95	120
***4	70	85	90	90	105	115	120	145
***3	80	100	105	105	120	130	145	170
***2	95	115	120	120	135	145	165	195
***1	110	130	140	140	160	170	190	220
***0	125	150	155	155	190	200	225	250
***00	145	175	185	185	215	230	250	280
000	165	200	210	210	245	265	285	315
0000	195	230	235	235	275	310	340	370
250	215	255	270	270	315	335	....	....
300	240	285	300	300	345	380	....	....
350	260	310	325	325	390	420	....	....
400	280	335	360	360	420	450	....	....
500	320	380	405	405	470	500	....	....
600	355	420	455	455	525	545	....	....
700	385	460	490	490	560	600	....	....
750	400	475	500	500	580	620	....	....
800	410	490	515	515	600	640	....	....
900	435	520	555	555	....	....	....	....
1000	455	545	585	585	680	730	....	....
1250	495	590	645	645	....	....	....	....
1500	520	625	700	700	785	....	....	....
1750	545	650	735	735	....	....	....	....
2000	560	665	775	775	840	....	....	....

Figure 4–1c ALLOWABLE AMPACITIES OF INSULATED COPPER WIRES (NEC TABLE 310–12).

Example 4–3

Find the ampacity of a type TW, No. 2/0 copper conductor:

a   In free air

b   Enclosed in a conduit with two other conductors of the same size and type

c   Enclosed in a conduit with 8 other conductors of the same type and size

Solution

a   From Table 310–13, the ampacity is 225 A.

b   From Table 310–12, the ampacity is 145 A.

c   From Table 4–1, I = 0.7 × 145 A = 101.5 A.

4–1   TEMPORARY WIRING

Temporary electrical power and lighting installations may be required during the period of construction, remodeling, or demolition of buildings or for a period not to exceed 90 days for Christmas decorative lighting, carnivals, and similar purposes. When installing temporary wiring for any of these purposes, electric services must be installed to conform with Article 230 of the NEC. This means that all requirements for a permanent service also apply to temporary wiring.

Feeders to subpanels and equipment must be provided with overcurrent protection (Article 240) and must originate in an approved distribution center. For temporary wiring, however, the conductors need not be enclosed in conduit but may be contained within multiconductor cords or cables or be run as open conductors on insulators not more than 10 feet apart where subject to mechanical injury.

Branch circuits, like feeders, must originate in an approved distribution panelboard and may be run to the various outlets in multicord or multi-cable assemblies. All conductors must be protected by overcurrent devices at their rated ampacity, and all cables should contain a separate equipment-grounding conductor.

All receptacles must be of the grounding type and be electrically connected to the grounding conductor. For receptacles installed on construction sites, an approved ground-fault circuit interrupter must be provided for the protection of personnel.

A detail drawing of a temporary electrical service and related wiring for a construction site is shown in Figure 4–2. This drawing shows the basic components of the service, with corresponding NEC articles for reference.

4–2   NONMETALLIC-SHEATHED CABLE

Nonmetallic-sheathed cable, shown in Figure 4–3, is called NM, Romex, or similar trade names in the electrical construction industry. Type NM cable (NEC Article 336) is used extensively for both exposed and concealed wiring in residential buildings in normally dry locations where it will not be exposed to dampness. Therefore, it may *not* be embedded in masonry, concrete, or plaster.

Although type NM cable is sometimes used in certain commercial applications (department stores), it may not be used in commercial garages, in theaters and assembly halls holding 200 or more people, in motion picture studios, in storage battery rooms, in hoistways, or in any hazardous location. Besides the restrictions set forth in the NEC, some local ordinances do not permit the use of NM cable in *any* commercial building. Always check this thoroughly before using NM cable in any application other than residential.

Nonmetallic cable must be fastened, preferably with staples or cable straps, at intervals not over 4-1/2 ft., except where it is fished through an opening between outlets. When the cable must be installed across joists, studs, etc.,

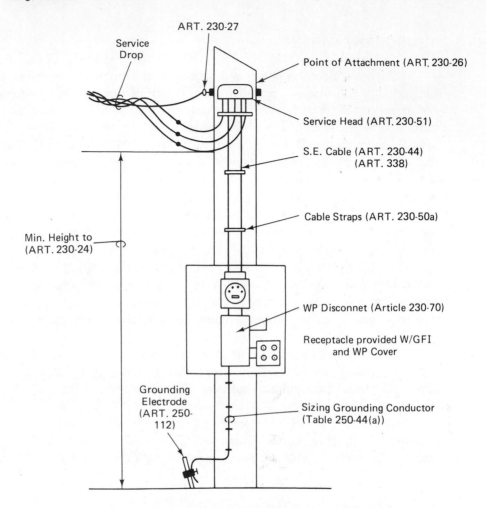

ART. 230-27

Service
Drop

Point of Attachment (ART. 230-26)

Service Head (ART. 230-51)

S.E. Cable (ART. 230-44)
(ART. 338)

Cable Straps (ART. 230-50a)

Min. Height to
(ART. 230-24)

WP Disconnet (Article 230-70)

Receptacle provided W/GFI
and WP Cover

Grounding
Electrode
(ART. 250-
112)

Sizing Grounding Conductor
(Table 250-44(a))

Figure 4–2   INSTALLATION DETAILS OF A TYPICAL TEMPORARY ELECTRIC
SERVICE FOR A BUILDING CONSTRUCTION SITE.

14-2G ANACONDA DUTRAX TYPE NM 600 V

Figure 4–3   NONMETALLIC SHEATHED CABLE; 14-2 WITH GROUND WIRE.

directly from one to another, the timbers must either be notched, or running
boards must be installed for mechanical protection. It is a good practice, where
possible, to run the cable through joists and studs or on the sides of the wood
members. It is also good practice to keep NM cables away from any areas where
nails are likely to be driven later, i.e., window and door frames.

Another type of nonmetallic cable for residential use is NMC. This type of cable may be used wherever NM is permitted and in moist, damp, or corrosive locations and outside or inside walls of masonry blocks or tile. Neither type NM or NMC can be used, however, as service-entrance cable, nor can either be embedded in poured concrete. Figure 4–4 shows NEC articles pertaining to installation of type NM and NMC cables.

Another cable in the NM family is type UF. It is suitable for use in any of the places where types NM and NMC are used. It is particularly intended for underground feeders (such as to an outside residential post lamp), including direct burial in the earth, if adequate protection is provided.

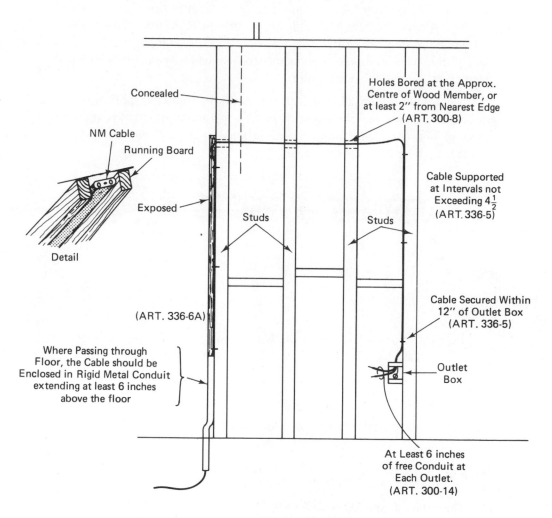

Figure 4–4   ILLUSTRATION SUMMARIZING THE USE OF TYPE-NM CABLE.

4–3   METAL-CLAD CABLE OR BX

Type MC (metal clad) or AC (armored cable) is widely used for wiring in both residential and small commercial buildings. It is installed in new work much the same as nonmetallic cable is (Figure 4–5). Its flexibility and compactness make it very useful for modernization and alteration work where the cable must be fished inside of finished partitions.

Type MC cable (Article 334 of the NEC) is a fabricated assembly of two, three, or four conductors, ranging in size from No. 14 AWG to a maximum of 500 MCM, in a flexible metallic enclosure (see Figure 4–5). Over the years, the trade name ''BX'' has been used in most circles to identify this type of cable.

The wires in modern MC cable are covered with thermo-plastic (TW) insulation while the steel armor has a corrosion-resistant coating. Type ACL, a similar cable, contains lead-covered conductors for use where the cable is exposed to the weather or to continuous moisture.

A common BX installation fault is to bend the cable too short; this breaks the armor and exposes the conductor to damage or corrosion. Bends should have a radius of not less than 7 times the diameter of type MC cable nor 5 times the diameter of type AC or ACL cable.

Figure 4–5   TYPE MC (BX) CABLE.

Example 4–4

Type AC No. 14/2 (2-conductor, No. 14 cable, called ''fourteen-two'') has an outside diameter of 0.52 inches. Use the limitations given in the previous paragraph to calculate:

a   The radius of a 90° bend for type AC cable.

b   The radius of a 90° bend for type MC cable.

Solution

a   The radius of type AC cable for a 90° bend must be at least 5 times the diameter of the cable; therefore $5 \times 0.52 = 2.6$ inches = radius of bend.

b   The radius of type MC cable must be at least 7 times the diameter of the cable; therefore, $7 \times 0.52 = 3.64$ inches = radius of bend.

Exposed MC or AC cable should be fastened to building surfaces by staples or BX straps at points no more than 4-1/2 feet apart and within 12 inches of every outlet box or termination. Type MC or AC cable runs should follow the building surfaces in order to keep the cable from being damaged. If the cable is to be installed through holes bored in studs or joists, the holes should be bored in the middle of the wood members not less than 2 inches from the edge. At this distance, the cable is not likely to be damaged by nails. Where the cable is laid in notches in wood, steel plates 1/16-inch thick should be placed over the notches in order to prevent cable damage by nails.

Most electricians prefer MX or BX cable for work on existing homes where much fishing of the cables is required. It is just stiff enough to be pushed rather accurately through void spaces in partitions, yet flexible enough to bend around corners.

4-4   TYPE MI AND ALS CABLES

Mineral-insulated, metal-sheathed (Type MI) cable (Article 330 of the NEC) is a cable in which one or more electrical conductors are insulated with a highly compressed refractory mineral insulation and are enclosed in a liquidtight and gastight metallic tube sheathing. Special fittings are used for terminating and connecting the tubing to outlet boxes and other equipment. Its use in electrical construction work is almost unlimited because it is approved for use in exposed and concealed work, in dry or wet locations, in all classes of hazardous location, embedded in plaster finish, in underground runs, and in practically any other location.

Aluminum-sheathed (Type ALS) cable (Article 331) is a factory-assembled cable consisting of one or more insulated conductors enclosed in an impervious, continuous, closely fitting tube of aluminum. It must be used with approved fittings for terminating and connecting to outlet boxes and other equipment. This type of cable may be used in exposed and concealed work and in dry and wet locations. It is *not* approved for use in concrete or in areas subject to severe corrosive influences, unless suitable, supplemental corrosion protection is provided.

Both MI and ALS cables are installed in practically the same manner. They are usually installed in *continuous runs* from distribution points to the load without use of splices, pull boxes or joints. It is supplied either in self-supporting coils or wound off installer-devised reels (see Figure 4–6). It can be unwound to run along surfaces of brick, plaster, concrete, steel, or any other materials except those prohibited by the NEC. The cable is fastened in place by approved clamps and straps that match the sheath diameter and the number of cables in the run. The interval between supports should not exceed 6 feet, except where the cable is fished.

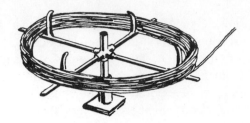

Figure 4–6   METHOD OF USING REELS TO INSTALL CONTINUOUS RUNS OF TYPE MI OR ALS CABLE.

A bending radius of not less than six times the cable diameter is recommended. This assures a good appearance and allows for straightening and rebending where necessary. Sharper bends are permissible, but subsequent straightening requires care, and annealing of the cable with a torch before straightening may be necessary to avoid ''rippling'' the sheath. Special bending hickeys with padded faces to prevent marking the sheath are available.

To strip, terminate, and seal the cable, strip the cable sheath with a special stripping tool (Figure 4–7a). A gland connector is then slipped on the cable before a screw-on self-tapping pot is installed (see Figure 4–7b). Continue by filling the pot with plastic sealing compound and slip the sleeving subassembly into position over the bored conductors (Figure 4–7c). Finally, use a crimping and compressing tool to compress and crimp the cap into the pot (Figure 4–7d).

4–5   SERVICE-ENTRANCE CABLE (TYPES SE AND USE)

Service-entrance (SE) cable looks similar to type NM cable. It has two or three insulated conductors around which a stranded, uninsulated neutral conductor is wound. Common sizes include two No. 2 insulated aluminum conductors and a No. 4 bare wire, 2/0, and 4/0

Service-entrance cables are used as service-entrance conductors on residential and small commercial buildings (NEC Article 338). The cable offers the advantage of being easily carried in coils; thus, only the length needed is cut off. The cable should be installed on buildings so that it will not be subject to mechanical damage.

For use in interior wiring systems, service-entrance cable may be used in the same fashion as NM cable if *all* the conductors of the cable are insulated. If the neutral or grounded conductor is bare, the cable may be used only for a circuit supplying an electric range, a clothes dryer, built-in cooking units, or other electric equipment where the uninsulated conductor is used only as an equipment ground.

Figure 4–7a STRIP THE CABLE SHEATH WITH A SPECIAL STRIPPING TOOL.

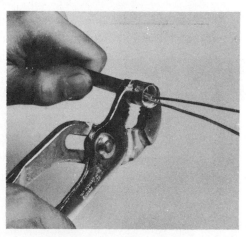

Figure 4–7b SLIP THE GLAND CONNECTOR ON THE CABLE BEFORE INSTALLING THE SCREW ON POT.

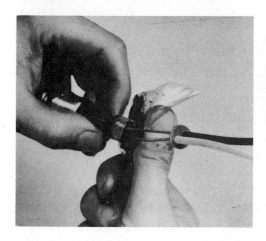

Figure 4–7c FILL THE POT WITH THE PLASTIC SEALING COMPOUND AND SLIP THE SLEEVING SUBASSEMBLY INTO POSITION.

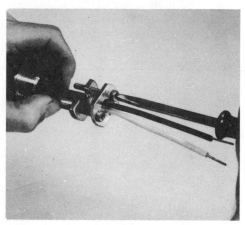

Figure 4–7d USE THE CRIMPING AND COMPRESSING TOOL TO COMPRESS THE CAP INTO THE POT.

4–6   RIGID METAL CONDUIT

The principal difference between cable wiring and rigid metal conduit (or any raceway) wiring (Article 346 of the NEC) is that the latter allows for conductors to be pulled in and pushed out of conduit or raceway system without affecting the system or the building structure. This provides the capability of either replacing or adding conductors without too much expense (as compared with the cost of installing an entirely new system). It is the recommended system for all industrial and most large commercial applications.

All raceway systems (rigid conduit, EMT, etc.) are completely installed and secured in or to the building structure before any wiring is installed in the raceway.

Rigid metal conduits (Figure 4–8) are probably the most widely used of the various raceway systems because they may be used under all atmospheric conditions and for all applications. Where subjected to severe corrosive areas, however, the rigid metal conduit may need a protective covering. This wiring system is well suited to withstanding hard usage encountered during construction of masonry buildings and hard industrial use. Rigid metal conduit is considered to be the finest system because it affords maximum protection of the conductors.

While there are several types of rigid metal conduits, standard galvanized is the most popular. This type comes in lengths of 10 feet, including one coupling, and is available in trade sizes from 1/2 to 6 inches; the sizes give the approximate inside dimensions. The range of sizes is shown in NEC Table 346–10a and 10b, as well as in NEC Chapter 9, Table 1.

Standard elbows, commonly called "ells" or "90-degree bends," are factory made in all sizes, but bends of 1 inch in diameter and smaller are usually formed on the job with hand benders or hickeys. The larger sizes may also be bent and formed on the job, but this requires the use of a mechanical bender, described in Chapter 9.

In general, no conduit smaller than 1/2-inch trade size would be used. All cut ends must be reamed to remove burrs and rough edges. Where a conduit enters a box or other fitting, a *bushing* is recommended to protect the wires from abrasion.

The number of conductors permitted in a single conduit must be calculated in accordance with the percentage fill specified in Chapter 9, Table 1 of the NEC. Pull boxes may be installed in extra-long runs of conduit. The NEC limits the number of 90° bends in any run to four, *or the equivalent;* that is, a single run of conduit between boxes might have two 90° bends and four 45° bends, but no more.

Rigid metal conduit is always installed as a complete system as provided in Article 300 of the NEC. The conduit should be firmly fastened within 3 feet of each outlet box, junction box, cabinet, or fitting. The conduit should also be supported at least every 10 feet, except straight runs of rigid conduit made up with approved threaded couplings, which may be secured according to Table 4–2. Only then are conductors pulled or fished through the conduit.

Figure 4–8   GALVANIZED RIGID METAL CONDUIT.

Table 4–2

Conduit size (inches)	Maximum distance between rigid metal conduit supports (feet)
½–¾	10
1	12
1¼–1½	14
2–2½	16
3 and larger	20

When threadless couplings and connectors are used with rigid conduit, they should be made tight. Where the conduit is to be buried in concrete, the couplings and connectors must be of the concrete type; where used in wet locations, they must be of the raintight type.

4–7   RIGID NONMETALLIC CONDUIT

Rigid nonmetallic (PVC) conduit (NET Article 347) is being used more and more each year in electrical installations. Besides requiring fewer tools, and in most cases less time, to install, it has the advantages of being lightweight and highly resistant to moisture and chemical atmospheres. Rigid, nonmetallic conduit may be used for direct burial in the earth; in walls, floors, and ceilings; in locations subject to severe corrosive influences; and in cinder fill and wet locations.

Rigid nonmetallic conduit should *not* be used in hazardous locations, except as covered in Articles 514–8 and 515–5 of the NEC. Furthermore, it shall not be used for the support of lighting fixtures or similar equipment or where subject to physical damage, unless approved for the purpose. It should not be used in areas where the ambient temperature exceeds that for which the conduit or the conductors have been approved. Nor should it be used to contain circuits whose potentials exceeds 600 volts, unless the conduit is encased in not less than 2 inches of concrete.

During the installation of rigid nonmetallic conduit, all cut ends must be trimmed inside and outside to remove any rough edges. All joints should then be made by a method specifically approved for the purpose—usually joined with fittings and cement.

The conduit must be supported within 4 feet of each termination point and according to Table 347–8 of the NEC: this table lists maximum spacing between supports for various sizes of nonmetallic conduit and for different ratings of conductors. Expansion joints must also be provided where required to compensate for thermal expansion and contraction. The number of conductors allowed in the various sizes of rigid nonmetallic conduit corresponds to Table 1, Chapter 9 of the NEC, the same as for rigid metal conduit.

Complete installation instructions for rigid nonmetallic conduits may be found in Chapter 8 of this book.

4–8 ELECTRICAL METALLIC TUBING

Electric metallic tubing (EMT) (NEC Article 348) may be used for both exposed and concealed work, *except* where it will be subjected to severe damage during use. Furthermore, it may not be used in cinder concrete or fill where subjected to permanent moisture, unless some means of protecting it is provided. However, the tubing may be installed a minimum of 18 inches under the fill.

Threadless couplings and connectors are used for EMT installations and should be installed so that the tubing will be made up tight. Where EMT is buried in masonry or is installed in wet locations, couplings and connectors as well as supports, bolts, straps, screws, etc., should be of a type approved for the condition.

Bends in EMT are made with a tubing bender so that no tubing injury will occur and so that the internal diameter of the tubing is not effectively reduced. Bends between outlets or termination points should contain no more than the equivalent of 4 quarter bends (360 degrees total), including those bends located immediately at the outlet or fitting (offsets).

All cuts in EMT are made with either a hack saw, power hack saw, tubing cutter, or other approved device. Once cut, the tubing ends should be reamed with a screwdriver shank or pipe reamer in order to remove all burrs and sharp edges that might damage the conductor insulation where the conductors are pulled in the conduit.

4–9 FLEXIBLE METAL CONDUIT

Flexible metal conduit (NEC Article 349) generally comes in two types: a standard metal-clad type and a liquidtight type. The former cannot be used in wet locations, unless the conductors within the conduit are of a type specially approved for such conditions. Neither type can be used where it will be subjected to physical damage or where ambient and/or conductor temperature will produce an operating temperature in excess of that for which the material is approved. Other uses are fully described in Articles 350 and 351 of the NEC.

When this type of conduit is installed, it should be secured by approved means at intervals not exceeding 4-1/2 feet and within 12 inches of every outlet box, fitting, or other termination points. However, exceptions exist: for example, when flexible metal conduit must be fished in walls, ceilings, etc., securing the conduit at these intervals would not be practical. Also, where more flexibility is required, lengths of not more than 3 feet may be utilized at termination points.

Flexible metal conduit may be used as a grounding means where both the conduit and the fittings are approved for the purpose. In lengths over 6 feet, it is best to install an extra grounding conductor within the conduit for added insurance.

4–10 SURFACE RACEWAYS

When it is impractical to install the wiring in concealed areas, surface metal molding or raceways (NEC Article 352) are a good compromise. Although metal molding is visible, proper painting to match the color of the ceiling and walls makes it very inconspicuous.

Surface raceways are made from sheet metal strips drawn into shape and come in various shapes and sizes with factory fittings to meet nearly every application found around the home. A few of the fittings available are shown in Figure 4–9.

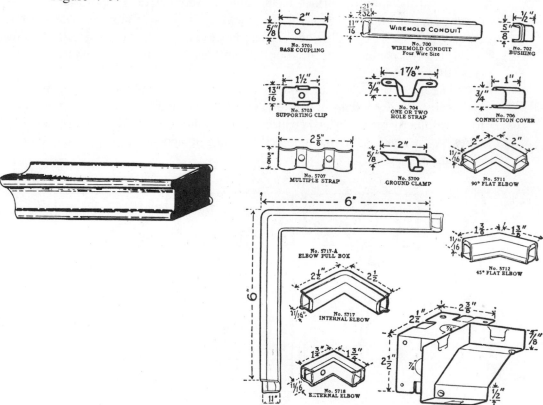

Figure 4–9 SURFACE METAL RACEWAY WITH VARIOUS FITTINGS.

Running straight lines of surface molding is simple. A coupling is slipped in the end of a length of molding, out enough so that the screw hold is exposed, and then the coupling is screwed to the surface to which the molding is to be attached. Another length of molding is then slipped on the coupling.

Factory fittings are used for corners and turns, or the molding may be bent (somewhat) with a special bender (Figure 4–10). Matching outlet boxes for surface mounting are also available, and bushings are necessary at such boxes in order to prevent the sharp edges of the molding from injuring the insulation on the wire (Figure 4–11).

Clips are used to fasten the molding in place, the clip is secured by a screw, and the molding is then slipped into the clip (Figure 4–12). Where extra support of the molding is needed, straps, fastened by screws, may be used. When parallel runs of molding are installed, they may be secured in place by means of a multiple strap. The joints in runs of molding are covered by slipping a connection cover over the joints. Such runs of molding should be grounded the same as any other metal raceway, and this is done by use of grounding clips. The current-carrying wires are normally pulled in after the molding is in place.

The installation of surface metal molding requires no special tools, unless bending the molding is necessary. The molding is fastened in place with screws, toggle bolts, etc., depending on the materials to which it is fastened. All molding should be run parallel with the room or building lines, i.e., base boards, trims, and other room molding.

The decor of the room should be considered first, and the molding made as inconspicuous as possible. It is often desirable to install surface molding not used for wires in order to complete a pattern set by molding containing current-carrying wires or to continue a run in order to make it appear as part of the room's decoration.

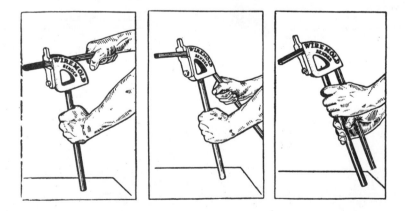

Figure 4–10 USE OF A SPECIAL SURFACE-METAL RACEWAY BENDER.

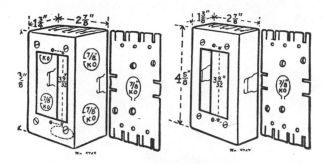

Figure 4–11   MATCHING OUTLET BOXES FOR THE SURFACE MOUNTING ARE AVAILABLE FOR USE WITH SURFACE METAL MOLDING.

Figure 4–12   CLIPS ARE NORMALLY USED TO FASTEN THE MOLDING IN PLACE.

4–11   UNDERFLOOR RACEWAYS

Underfloor ducts (NEC Article 354), either steel or fiber, are frequently used in the wiring system design for power and communications, particularly where a grid system for future service is desired. The ducts are installed in the rough floor slab prior to pouring the concrete and are supported to obtain the proper elevation below the finished slab; then the finished floor is poured.

Steel cellular floors are also a form of underfloor raceways. Such systems provide a maximum grid raceway system for present and future needs and a supporting structure for the concrete floor that is poured over the steel cellular floors.

Underfloor raceways (Figure 4–13) should not be used where subject to corrosive vapors nor in any hazardous location (Article 354–2). In cellular metal floor raceways, some of the cells are used as raceways for other than electrical circuits or equipment, such as steam, water, and gas pipes. No electric conductors can be installed in any cell or header that contains such mechanical pipes or equipment (Article 356–2).

In general, the total cross-sectional area of all conductors in underfloor raceways must not exceed 40 percent of the cross-sectional area of the header or

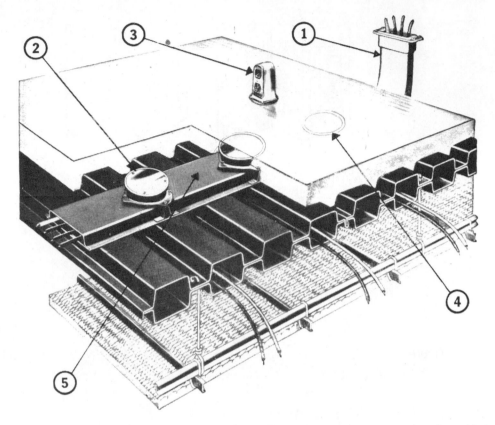

1. Vertical ell to extend the header to panel or cabinet.

2. Junction unit where the wires make a turn from the header into the floor cell.

3. Floor outlet for electrical service placed exactly where desired.

4. Floor covering adapter.

5. Standard header duct.

Figure 4–13   APPLICATION OF UNDERFLOOR WIRING SYSTEM.

cell in which they are located. Where the raceway contains only type AC, metal-clad cable or type NM cable, these requirements do not apply (Article 356–5).

A suitable number of markers must be installed during the construction of an underfloor raceway system for system identification and so that the cells may be located in the future and any floor or junction boxes used in the system should be leveled to the floor grade and sealed against the entrance of water. The boxes should be constructed of metal and be electrically continuous with the raceway.

4–12 BUSWAYS

Various types of bus duct systems with self-contained conductors of copper strap buses are manufactured for use as substitutes for conduit and wire or cable systems for specialized feeder and power installations, especially in industrial applications. Such systems are installed exposed with the proper hangers after the building structure is completed.

Busway systems are extremely useful in applications where there is a need for surface wiring that provides ready access to conductors in order to obtain power for electrical equipment that is subject to change at frequent intervals. A busway system also provides a mechanical protection around the conductors, as shown in Figure 4–13.

Busway systems may be installed only for exposed work, as pointed out in the NEC Article 364–2. Busways should also not be installed where they will be subject to severe physical damage or corrosive vapors, in hoistways, in any hazardous location, nor outdoors or in wet or damp locations unless specially approved for the purpose.

The system should be securely supported at intervals not exceeding 5 feet. Where the busway is installed in a vertical position, the supports for the bus bars must be designed for such an installation.

Branches from busways may be made with either conduit or other busways. When approved fittings are used, even metal-clad cable and suitable cords may be used.

Overcurrent protection should be provided, as specified in Articles 364–9 through 364–13 of the NEC. However, overcurrent protection may be omitted at points where busways are reduced in size, provided that the smaller busway does not extend more than 50 feet, has a current rating at least equal to one-third the rating or setting of the previous overcurrent device, and is free from contact with combustible material.

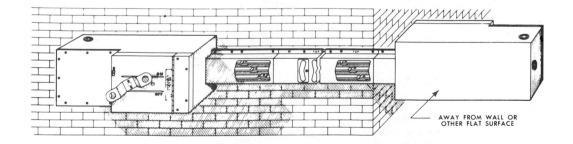

AWAY FROM WALL OR
OTHER FLAT SURFACE

Figure 4–14 APPLICATION OF A BUSWAY SYSTEM.

Busways that are used as branch circuits and are so designed that loads can be connected at any point shall be limited to lengths that will insure that the circuits will not be overloaded in normal use. A rule of thumb is the length of such runs in feet should not exceed 3 times the ampere rating of the branch circuit.

4–13   CABLE TRAY SYSTEMS

Cable trays are the usual means of supporting cable systems in industrial applications. The structure is made of metal or noncombustible units forming a continuous rigid assembly for carrying electrical cables from their origin to their point of termination, frequently over considerable distances (see Figure 4-14).

Cables rest upon the bottom of the tray and are held in the tray by two longitudinal side rails. Once the cable in the tray reaches its destination, a conduit is used, in most cases, to finish the run from the channel to the actual termination.

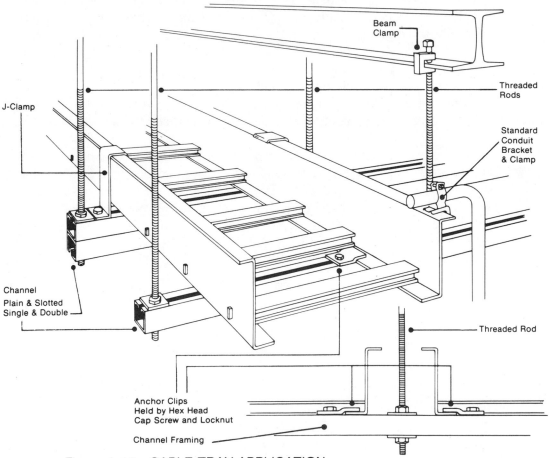

Figure 4–15   CABLE TRAY APPLICATION.

Cable tray systems (defined in NEC as "continuous rigid cable supports") may be used as the mechanical support only for the following wiring methods under the conditions detailed in the article for each wiring method:

1   Mineral-insulated, metal-sheathed cables

2   Aluminum-sheathed cable

3   Metal-clad cable (Article 334)

4   Nonmetallic-sheathed cable (Article 336)

5   Multiple-conductor, service-entrance cables (Article 339)

6   Multiple-conductor, underground feeder (UF) cable (Article 339)

7   Any approved conduit or raceway with its contained conductors

8   Shielded nonmetallic-sheathed cable (Type SNM) for hazardous locations (Article 337)

Cable trays (Figure 4–15) may also be used as the mechanical support for factory-assembled, multiconductor control, signal, and power cables that are specifically approved for installation in continuous rigid cable supports and in fire-resistive or noncombustible construction. However, the system should not be used in hoistways where the supported cables are subject to severe physical damage, in areas having readily combustible contents, and in similar locations.

When cable tray systems are used, all the metal sections of the system, including fittings and supports, must be bonded and effectively grounded in order to provide a continuous circuit for fault current. It cannot, however, be used as a grounded circuit conductor or as an equipment grounding conductor, as specified in Article 250–33 of the NEC.

QUESTIONS

Answer the following questions by filling in the blanks.

1   All conductors of No. 6 AWG and larger used in a raceway system must be _____ , not solid wire.

2   Type SE cable must be used to feed a subpanel within a building only if the neutral wire is _____ .

3   The smallest wire allowed by the NEC for branch circuit wiring is size _____ AWG.

4   Type AC, metal-clad cable may be used to wire an assembly hall only if the seating capacity is less than _____ persons.

5   The minimum distance to the first staple, strap, or other support on exposed type NM cable is _____ inches from the outlet box.

6   Service-entrance cable may be used on the outside of a building for single-phase service-entrance conductors and may also be used to feed an electric _____ and _____ inside the building, even with an uninsulated neutral.

7   The installation of a rigid metal conduit system should be _____ before conductors are pulled in the system.

8   _____ are used primarily on type AC (BX) cable to prevent the conductors from becoming damaged.

9   The minimum size of flexible metallic conduit that may be used to supply a branch circuit is _____.

10   The article in the NEC that prohibits the use of running threads on rigid metallic conduit is Article _____.

11   _____ rigid conduit may be installed directly in a cinder fill.

12   Armored cable containing _____ conductors may be used in damp locations.

13   Bends in flexible metallic conduit must have a radius of at least _____ times the diameter of the cable.

14   The minimum radius for a bend of 2-inch rigid metallic conduit used to contain lead-covered cable is _____.

15   Armored cable, even those containing lead-covered conductors, may not be used in _____ atmospheres.

16   The maximum distance between staples or other supports holding type AC (BX) cable is _____ feet.

17   Three locations in which EMT may not be used are _____, _____ and _____.

18   Three types of cable suitable for use in cable tray systems are _____, _____ and _____.

19   When smaller sizes of busduct are connected to larger sizes, they need not be used unless the run is over _____ feet.

20   One type of cable suitable for direct burial in the earth (provided adequate protection is provided) is called _____.

PROBLEMS

1   A solid wire has a diameter of 0.6325 inches. Calculate the diameter in mils.

2   The circular mil area of the wire described in Problem 1 is _____ CM.

3   A wire with an area of 750,000 CM has a diameter of _____ mils.

4   The answer obtained in Problem 3 is equivalent to _____ inches.

5   Find the ampacity of type THW wire, No. 4/0 copper conductor enclosed in a conduit with three other conductors of the same size. One of these conductors is a neutral conductor and carries only the unbalanced load of the other three conductors.
HINT: Use Table 310–12 in the NEC.

5

ELECTRICAL CALCULATIONS

Since most problems encountered in the design and layout of electrical systems involve the use of equations and calculations, it is essential that all electrical personnel have a good understanding of electrical equations and how to apply them to actual on-the-job applications. The equations and solutions contained in this chapter are arranged in a logical sequence for easy reference and the combined total should cover most of the problems that the electrician will face in the electrical, building-construction industry. The electrical designer will also find these equations helpful for most problems encountered in design and layout.

5–1   BASIC ELECTRICAL CALCULATIONS

George S. Ohm, a German physicist, discovered that the current through an electrical conductor depends on the amount of pressure (volts) and the resistance of the circuit components. These laws or equations are summarized in Figure 5–1. They are directly applicable to any resistive circuit, any resistive portion of a circuit, any d.c. circuit, and any a.c. circuit or portion of an a.c. circuit with a power factor of 100 percent.

In general, Ohm's Law states that the current, in amperes, varies directly with the pressure difference, in volts. It further states that doubling the resistance will permit only half as much current to flow in the circuit; cutting the resistance in half will permit twice as much current to flow. Stated another way, the current increases proportionately to every decrease in resistance, while the current decreases proportionately to any increase—provided the voltage remains the same throughout.

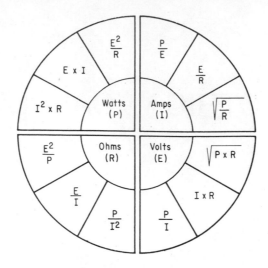

Figure 5–1   SUMMARY OF OHM'S LAW.

A review of the three basic electrical units—ampere, ohm, and volt—is in order before work is done with Ohm's Law. An ampere is an electrical unit used to measure the flow of current in a circuit. The resistance or opposition to the flow of current is measured in ohms. The external force applied to a circuit to overcome the opposition to the flow of current is measured in volts. The relationship among these units (Ohm's Law) is one of the most important laws used in electrical calculations, and every electrician should be familiar with the law and its application to practical electrical problems.

The three basic ways of stating Ohm's Law are listed below, where I = amperes, R = ohms, and E = volts:

1    $E = IR$; the voltage equals the current multiplied by the resistance.

2    $I = \dfrac{E}{R}$; the current equals the voltage divided by the resistance.

3    $R = \dfrac{E}{R}$; the resistance equals the voltage divided by the current.

The electrical unit for power, the watt, may be incorporated into Ohm's Law for further calculations. Where W = watts, current may be found by the equations:

1    $I = \dfrac{W}{E}$; the wattage divided by the voltage equals current.

2    $I = \sqrt{\dfrac{W}{R}}$; the square root of the wattage divided by the resistance equals current.

Voltage may be found by the equations:

1    $E = \dfrac{W}{I}$; wattage divided by the current equals volts.

2    $E = \sqrt{WR}$; the square root of the wattage times the resistance equals voltage.

Resistance may be found by the equations:

1    $R = \dfrac{E^2}{W}$; the voltage squared divided by the wattage equals resistance.

2    $R = \dfrac{W}{I^2}$; the wattage divided by the current squared equals the resistance.

The power, in watts, of a circuit may be found by the following equations:

1    $W = \dfrac{E^2}{R}$; the voltage squared divided by the resistance equals watts.

2    $W = I^2 \times R$; the current squared times the resistance equals watts.

3    $W = EI$; the voltage times the current equals watts.

Conductor Areas

Much information concerning conductors and their resistance and current-carrying capacity can be obtained from convenient tables (see Appendix). These tables should be used whenever possible. However, when tables are not available, a knowledge of simple wire calculations becomes very important. For example, tables in the NEC give the allowable current-carrying capacities of various conductors based on the heating of conductors but does not consider voltage drop due to resistance of long runs. Both of these considerations are very important and should always be kept in mind when the conductors of any wiring system are sized.

Conductor sizes are commonly specified in American Wire Gauge (AWG). Gauge numbers are arranged according to the resistance of wires: the larger numbers are for the wires of greatest resistance and smallest area. The most common sizes for lighting and power wiring for building construction are from 4/0 (four ought) down to No. 14; sizes down to No. 22 are used for signal and communication wiring.

In addition to the AWG numbers a unit called the *mil* is sued for measuring the diameter and area of conductors. A mil is equal to 0.001 inch. For round conductors, the *circular mil* unit, which is the area of the circle with a diameter of 0.001 inch is used. The abbreviation commonly used for circular mil is CM. To obtain the area of a round conductor in circular mils, the diameter (in mils) is squared:

$$CM = D^2$$

In many cases, it becomes necessary to calculate the resistance of a certain length of wire of a given size. If the unit resistance of the conductor metal is known, this calculation is very simple. For example, the resistance of ordinary copper is 10.79 ohms per mil foot; for practical purposes, 11 ohms will suffice. Aluminum is approximately twice this figure and 22 ohms will suffice for all practical purposes. These figures are very important and should be remembered: they are "constants" used in some voltage-drop equations.

Effective Resistance

The resistance of an electrical circuit or load is of little use to the practical workmen, except as a step towards finding other electrical values more useful in electrical installations. The rule most often used to find the total resistance in a series circuit is:

THE TOTAL RESISTANCE IN A SERIES CIRCUIT IS EQUAL
TO THE SUM OF ALL RESISTORS IN THE CIRCUIT:

$$R_t = R_1 + R_2 + R_3 + \text{etc.}$$

Most circuits used in electrical systems for building construction are connected in parallel, and the total resistance in these circuits must sometimes be found in order to determine the circuit drawn by or the power expended in the circuit. The rule most often used to find the total resistance in a parallel circuit is:

THE SUM OF THE RECIPROCALS OF THE SEPARATE RESISTORS EQUALS THE
RECIPROCAL OF THE EQUIVALENT RESISTANCE:

$$1/R_t = 1/R_1 + 1/R_2 + 1/R_3 + \text{etc.}$$

Impedance

Most electrical systems for light and power encountered in building construction contain inductance. In circuits with only pure resistance loads, the inductance is usually small enough to be neglected in circuit calculations. However, in circuits with motors, relays, transformers, electrical discharge lighting, etc., the inductance may be significant enough to be included in the circuit calculations.

Current through an induction lags the voltage by 90°. Resistance and the inductive reactance (X_L), which provide opposition to the flow of current, may also be thought of as being 90° apart. The total opposition to the flow of current is called impedance and may be represented by the hypotenuse of the triangle formed as shown in Figure 5–2 and may be solved with the principles of trigonometry. Impedance, like other oppositions to the flow of current, is expressed in ohms.

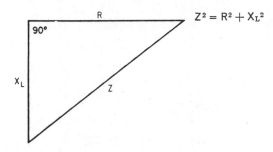

Figure 5–2 DIAGRAM SHOWING HOW IMPEDANCE IS EXPRESSED IN TRIGONOMETRIC FUNCTIONS.

When the resistance and inductive reactance are known in a circuit, the impedance may be found by the equation:

$$Z = \sqrt{R^2 + X_L^2}$$

In a.c. circuits, impedance is used in Ohn's Law just like resistance:

$$E = IZ$$

$$I = \frac{E}{Z}$$

$$Z = \frac{E}{I}$$

In circuits with both inductive reactance and capacitive reactance, the equation for finding impedance is:

$$Z = \sqrt{R^2 + (X_L - X_C)}$$

where, Z = impedance, R = resistance, X_L = inductive reactance, and X_C = capacitive reactance.

5–2   VOLTAGE DROP

In all electrical wiring, the conductors should be sized so that the voltage drop never exceeds 3 percent for power, heating, and lighting loads or any combinations thereof. Furthermore, the maximum total voltage drop for conductors, feeders, and branch circuits should never exceed 5 percent overall.

The voltage drop in any two-wire, single-phase circuit consisting mostly of resistance-type loads with negligible inductance may be found by the following equation:

$$VD = \frac{2K \times L \times I}{CM}$$

where VD = drop in circuit voltage, R = resistance per foot of conductor, I = current in the circuit, CM = area of conductor in circular mils, and K = resistivity of conductor material.
L = length of circuit (both ways) in feet

With this equation, the voltage drop in a circuit consisting of No. 10 AWG wire, 50 feet in length, and carrying a load of 20 amperes is:

$$VD = \frac{2 \text{ (in equation)} \times 50 \text{ (length)} \times 20 \text{ (current)} \times 11 \ (K)}{10,380 \text{ (area in CM of No. 10 wire)}}$$

The answer is 2.11 volts.

5–3   LOADING OF BRANCH CIRCUITS

The NEC specifically states that the total load on a branch circuit, other than motor loads, must not exceed 80 percent of the circuit rating when the load constitutes a continuous, i.e., a load that is in operation for 3 hours or longer.

The total load permitted on a 20-ampere circuit (No. 12 AWG fused at 20 amperes) that will be in operation for over 3 hours is 0.80 × 20 (amperes) = 16—the maximum permissible load. However, when circuits for most electrical outlets, especially lighting are being layed out, it is preferable to deal with watts rather than amperes. Assume that the circuit is supplied wtih 120 volts; use the equation P (watts) = E (volts) × I (current) or 120 × 16 = 1920. If the electrician were installing 100-watt lamps in fixtures fed by the circuit, a total of 9 fixtures could be connected.

It is a good practice to limit on new construction the load on a 20-ampere circuit to a maximum of 1200 to 1600 watts. This permits some additional loads to be added in the future, and it keeps the temperature of the conductors lower for better performance.

5–4   SIZING CONDUCTORS

Oftentimes, electricians are faced with the problem of sizing conductors for special electronic and electrical equipment on which conventional wire-sizing tables cannot be used. A typical examples follows.

An electronic, solid-state computer is located 460 feet from the main distribution panelboard. Since this type of equipment is very sensitive to voltage drops below that for which it is rated, the voltage drop must be kept to a minimum—less than the allowable 3 percent. The service consists of a three-phase, four-wire, 120/208-volt service, but the equipment is designed for use on a two-wire, 208-volt circuit. The allowable voltage drop is 2 percent, and copper wire will be utilized. To find the wire size that will carry the load with less than 2 percent voltage drop, use the equation:

$$CM = \frac{\text{length} \times \text{amperes} \times 2K}{\text{volts lost}}$$

In this equation, the number of feet must be measured or scaled one way— not both sides of the circuit. Volts lost should be taken as the drop allowed in volts—not the percentage. Circular mils show the size wire in AWG to use. In this case, $K = 11$ for copper.

There are only two unknowns in this case, the circular mils and the volts lost. To solve for the volts lost or voltage drop, multiply 208 (the voltage between phases) by 0.02 (the percentage of voltage drop permitted for the particular piece of equipment). The answer will be 4.16 volts. The nameplate on the piece of equipment gives a full-load ampere rating of 87 amperes at 208 volts. Substituting all known values into the equation, we have:

$$CM = \frac{460 \times 87 \times 22}{4.16} = 211{,}644 \text{ CM}$$

In the table in Figure 5–3, 250,000 CM (250 MCM) is the closest wire size normally available and is therefore the size to use for the circuit feeding this piece of equipment.

Size AWG MCM	Area Cir. Mils	Concentric Lay Stranded Conductors		Bare Conductors		D.C. Resistance Ohms/M Ft. At 25°C. 77°F.		
		No. Wires	Diam. Each Wire Inches	Diam. Inches	*Area Sq. Inches	Copper		Alumi-num
						Bare Cond.	Tin'd. Cond.	
18	1620	Solid	.0403	.0403	.0013	6.51	6.79	10.7
16	2580	Solid	.0508	.0508	.0020	4.10	4.26	6.72
14	4110	Solid	.0641	.0641	.0032	2.57	2.68	4.22
12	6530	Solid	.0808	.0808	.0051	1.62	1.68	2.66
10	10380	Solid	.1019	.1019	.0081	1.018	1.06	1.67
8	16510	Solid	.1285	.1285	.0130	.6404	.659	1.05
6	26240	7	.0612	.184	.027	.410	.427	.674
4	41740	7	.0772	.232	.042	.259	.269	.424
3	52620	7	.0867	.260	.053	.205	.213	.336
2	66360	7	.0974	.292	.067	.162	.169	.266
1	83690	19	.0664	.332	.087	.129	.134	.211
0	105600	19	.0745	.372	.109	.102	.106	.168
00	133100	19	.0837	.418	.137	.0811	.0843	.133
000	167800	19	.0940	.470	.173	.0642	.0668	.105
0000	211600	19	.1055	.528	.219	.0509	.0525	.0836
250	250000	37	.0822	.575	.260	.0431	.0449	.0708
300	300000	37	.0900	.630	.312	.0360	.0374	.0590
350	350000	37	.0973	.681	.364	.0308	.0320	.0505
400	400000	37	.1040	.728	.416	.0270	.0278	.0442
500	500000	37	.1162	.813	.519	.0216	.0222	.0354
600	600000	61	.0992	.893	.626	.0180	.0187	.0295
700	700000	61	.1071	.964	.730	.0154	.0159	.0253
750	750000	61	.1109	.998	.782	.0144	.0148	.0236
800	800000	61	.1145	1.030	.833	.0135	.0139	.0221
900	900000	61	.1215	1.090	.933	.0120	.0123	.0197
1000	1000000	61	.1280	1.150	1.039	.0108	.0111	.0177
1250	1250000	91	.1172	1.289	1.305	.00863	.00888	.0142
1500	1500000	91	.1284	1.410	1.561	.00719	.00740	.0118
1750	1750000	127	.1174	1.526	1.829	.00616	.00634	.0101
2000	2000000	127	.1255	1.630	2.087	.00539	.00555	.00885

Figure 5–3

5–5 SHORT-CIRCUIT CALCULATIONS

Several articles of the NEC require that adequate interrupting capacity and protection be provided for all components on any electrical system. To provide this protection, the short-circuit currents at various intervals within the electrical distribution system must be calculated. The steps necessary for the calculation (assuming an unlimited primary short circuit) are as follows:

1   Determine transformer full-load amperes from manufacturer's data, nameplate, or the following equations.

   a. Three-phase transformer:

$$\text{Current} = \frac{\text{kVA} \times 1000}{\text{line-to-line voltage} \times 1.73}$$

   b. Single-phase transformer:

$$\text{Current} = \frac{\text{kVA} \times 1000}{\text{line-to-line voltage}}$$

2   Find transformer multiplier by the equation:

$$\text{Multiplier} = \frac{100}{\text{Transformer \% impedance}}$$

3   Determine transformer ''let-thru'' short-circuit current from tales or by the equation:

Isca = transformer full-load amperes × multiplier found in step 2.

5–6 POWER FACTOR

The power factor of an alternating current is the number by which the apparent power in the circuit (volts times amperes) must be multiplied in order to ascertain true power. When an a.c. circuit contains inductance, the current lags behind the voltage; when it contains capacity, the current rises ahead of the voltage. In each case, the current and voltage reach their maximum values at different times, and the product of the current and voltage at any instant is less than it would be if the two were in phase with each other. If the voltage and current were measured separately, the voltmeter and ammeter would give the individual mean effective values; if they are measured by a wattmeter, the instrument indicates their combined effect synchronously, not the product of their effective values, which occur at different instants. Consequently, the wattmeter indication will be less than the product of the separate voltmeter and ammeter readings. The ratio of the power to this product is the power factor of the circuit. Expressed as a formula

$$\text{Power Factor} = \frac{\text{Watts}}{\text{Amperes} \times \text{Volts}}$$

This gives rise to the two methods of rating electrical apparatus, one on the basis of watts or kilowatts and the other on the basis of volt-amperes or kilowatt-amperes. The former represents actual power, usually in kW, while the latter represents the apparent power, usually in kVA (kilovolt amperes), generated, transmitted, or used by the apparatus. The latter rating is coming into more general use since it represents more adequately the voltage and current conditions to which the apparatus is subjected.

To find the current per phase in various sytems, use the following formulas:

$$I = \frac{W}{E \times P.F.} \quad \text{for single-phase circuit}$$

$$I = 0.50 \times \frac{W}{E \times P.F.} \quad \text{for two-phase circuit}$$

$$I = 0.58 \times \frac{W}{E \times P.F.} \quad \text{for three-phase circuit}$$

$$\text{Temp. °C} = {}^5/_9 \text{ (Temp. °F } -32)$$

$$\text{Temp. °F} = {}^9/_5 \text{ Temp. °C } -32$$

Where I = current in line in amperes, W = energy delivered in watts, E = potential between mains in volts, and P.F. = power factor. When the power factor cannot be accurately determined, it may be assumed as follows: lighting load with no motors, 0.95; lighting and motors, 0.85; motors only, 0.80.

QUESTIONS

Answer the following questions by filling in the blanks.

1   Ohm's Law states that the current varies directly with the pressure difference in
_____ .

2   Ohm's Law states that the current increases proportionately to every _____ of resistance.

3   The three basic ways of stating Ohm's Law are: _____ = amperes, _____ = ohms, and _____ = volts.

4   A 1500-watt electric baseboard heater operating on a 240-volt circuit will draw _____ amperes of current.

5   Conductor sizes are commonly specified in _____ _____ Gauge.

6   The area of a conductor with a diameter of 0.025 mils is _____ circular mils.

7   A series circuit containing three resistors of three ohms, 12 ohms, and 25 ohms, respectively will have a total resistance of _____ ohms.

8   An electrical circuit having two resistors connected in parallel rated at 15 ohms and 30 ohms will have a total resistance of _____ ohms.

9   The total opposition to the flow of electric current is called _____.

10   A 240-volt, two-wire, single-phase circuit has a total length (one way) of 125 feet. No. 10 AWG copper wire is used to feed a motor drawing 23 amperes. The voltage drop in this circuit will be _____ volts.

6

BRANCH CIRCUIT LAYOUT FOR LIGHTING AND POWER

The wires or other conductors that extend to the various outlets (lighting and power) from the panelboards are commonly referred to as branch circuits. From the standpoint of the NEC, the branch circuit is "the circuit conductors between the final overcurrent device protecting the circuit and the outlet(s) . . . a branch circuit supplying energy to one or more outlets to which appliances are to be connected; such circuits to have no permanently connected lighting fixtures not a part of an appliance . . . a branch circuit that supplies a number of outlets for lighting and appliances . . . a branch circuit that supplies only one utilization equipment . . . a branch circuit consisting of two or more ungrounded conductors having a potential difference between them, and an identified grounded conductor having equal potential difference between it and each ungrounded conductor of the circuit and which is connected to the neutral conductor of the system."

The size of the branch circuit wires or conductors varies, depending on the current requirements of the equipment connected to the outlet. Other factors, some of which are NEC requirements and some of which provide efficiency of operation over and above the NEC requirements, affect the wire or conductor sizes.

Branch circuit wires and conductors may vary from No. 14 copper wire to large-size, bare copper, rectangular buses. The aluminum conductor has been made available in recent years, and widespread use of aluminum for all sizes of building wires as well as for larger-sized cables is now in progress. Since the conductivity of aluminum is only about four-fifths that of copper, a larger aluminum conductor must be used to provide the same circuit capacity. This introduces some mechanical problems of making connections to the present type of wiring device and fixture terminals as well as the problem of electrical contact between dissimilar metals.

A simple branch circuit requires two wires or conductors to provide a continuous path for the flow of electric current. The usual lighting and convenience receptacle branch circuit operates at a nominal voltage or pressure of 120 volts. It is a requirement of the NEC that one of the conductors of a 120-volt circuit be grounded—usually to a neutral block at a panelboard. This branch circuit conductor must be identified by a white insulation or covering and is referred to as the neutral or grounded conductor. The ungrounded conductor is connected to the terminal of the overcurrent protective device.

In all but the very simplest two-wire wiring system, the neutral conductor may be common to two or more branch circuits, depending on the type of service supplied by the power company or by the owner's generating equipment. Because of this, two branch circuits could consist of two ungrounded conductors and a neutral conductor, a total of three; three branch circuits could consist of three ungrounded conductors and a neutral conductor, a total of four. These are commonly referred to as three-wire and four-wire branch circuits.

6–1 CONVENIENCE OUTLETS IN RESIDENTIAL OCCUPANCIES

Article 210–25 of the NEC specifically states the minimum requirement for the location of receptacles in residential buildings:

In every kitchen, family room, dining room, breakfast room, living room, parlor, library, den, sun room, bedroom, recreation room, or similar room, receptacle outlets shall be installed so that no point along the floor line in any wall space is more than 6 feet, measured horizontally, from an outlet in that space, including any wall space 2 feet or more in width and the wall space occupied by sliding panels in exterior walls. The wall space afforded by mixed room dividers, such as free-standing bar-type counters, shall be included in the 6-foot measurement.

In kitchen and dining areas, a receptacle outlet shall be installed at each counter space wider than 12 inches. Counter top spaces separated by range tops, refrigerators, or sinks shall be considered as separate counter top spaces. Receptacles rendered inaccessible by appliances fastened in place or appliances occupying dedicated space shall not be considered as these required outlets.

Receptacle outlets shall, insofar as practicable, be spaced equal distances apart. Receptacle outlets in floors shall not be counted as part of the required number of receptacle outlets unless located close to the wall.

At least one wall receptacle outlet shall be installed in the bathroom adjacent to the basin location.

For a one-family dwelling, at least one receptacle outlet shall be installed outdoors.

For a one-family dwelling, at least one receptacle outlet in addition to any provided for laundry equipment, shall be installed in each basement and in each attached garage.

Outlets in other sections of the dwelling unit for special applicances, such as laundry equipment, shall be placed within 6 feet of the intended location of the appliance.

At least one receptacle outlet shall be installed for the laundry.

As used in this section, a "wall space" shall be considered a wall unbroken along the floor line by doorways, fireplaces, and similar openings. Each wall space two or more feet wide shall be treated individually and separately from other wall spaces within the room. A wall space shall be permitted to include two or more walls of a room (around corners) where unbroken at the floor line.

The purpose of this equipment is to minimize the use of cords across doorways, fireplaces, and similar openings.

The receptacle outlets required by this section shall be in addition to any receptacle that is part of any lighting fixture or appliance, located within cabinets or cupboards, or located over 5-1/2 feet above the floor.

6–2   RECEPTACLES IN COMMERCIAL BUILDINGS

The only location requirement for commercial application of duplex receptacles is in show windows. In such areas, at least one receptacle outlet must be installed directly above a show window for each 12 linear feet or major fraction thereof of show window area measured horizontally at its maximum width.

All receptacles installed on 15- and 20-ampere branch circuits must be of the grounding type. Each receptacle must be calculated for a minimum load of 180 watts. Therefore, a conventional duplex receptacle must be calculated at 360 watts or more when designing branch circuits.

All receptacles installed for the attachment of portable cords must be rated at not less than 15 amperes. Metal faceplates shall be of ferrous metal not less than 0.030 inch in thickness or of nonferrous metal not less than 0.040 inch in thickness. Metal faceplates shall be grounded. Faceplates of insulating material shall be noncombustible and not less than 0.10 inch in thickness but shall be permitted to be less than 0.10 inch in thickness if formed or reinforced to provide adequate mechanical strength.

After installation, receptacle faces shall be flush with or project from faceplates of insulating material and shall project a minimum of 0.015 inch from metal faceplates. Faceplates shall be installed so as to seat against mounting surfaces. Boxes shall be installed in accordance with Section 370–10 of the NEC.

6–3   RECEPTACLES IN INDUSTRIAL LOCATIONS

In general, NEC requirements applying to the installation and use of receptacles in commercial applications also apply to industrial occupancies. However, there are a few cases where additional provisions apply. For example, receptacles and attachment plugs used in hazardous locations must be of the type providing for connection to the grounding conductor of a flexible cord and must be approved for the type location in which it is used.

Where condensed vapors or liquids may collect on, or come in contact with, the insulation on conductors, such insulation shall be of a type approved for use under such conditions; or the insulation shall be protected by a sheath of lead or by other approved means.

6–4   LOADING OF BRANCH CIRCUITS

The National Electrical Code specifically states that the total load on a branch circuit—other than motor loads—must not exceed 80 percent of the circuit rating when the load will constitute a continuous; that is, a load that is in operation for 3 hours or longer.

To size the total load permitted on a 20-ampere circuit (No. 12 AWG fused at 20 amperes) that will be in operation for 3 hours or longer, .80 × 20 (amperes) = 16—the maximum permissible load.

However, when laying out circuits for most electrical outlets—especially lighting—it is more desirable to deal with watts rather than amperes. Assume that the circuit is supplied with 120 volts; then use the equation P (watts) = E (volts) × I (current) or 120 × 16 = 1920. If the electrician was installing 100-watt lamps in fixtures fed by the circuit, a total of 9 fixtures could be connected.

It is a good practice to limit the load on 20-ampere circuits to a maximum of 1200 to 1600 on new construction. This permits some additional loads to be added in the future and it also keeps the temperature of the conductors lower for better performance.

6–5   LIGHTING OUTLETS

There is no general rule for locating lighting outlets for general illumination, but it is usually desirable to locate lighting fixtures so that the illumination in a given area is uniform. Where the number and location of lighting outlets cannot be determined from the architectural or electrical drawings, or by the arrangement of the decor layout, it is usually desirable to arrange the lighting outlets in the form of squares or rectangles (See Figure 6–1). In doing so, the outlets should be placed at the centers of the squares and not at the corners as shown in Figure 6–2. The latter arrangement will provide most of the light near the center of the room and will therefore create a very low intensity of illumination near the walls.

Examples of several layout patterns are shown in Figures 6–3, 6–4, 6–5, and 6–6. These floor plans should be studied—and details noted—for future lighting layout problems encountered by the reader.

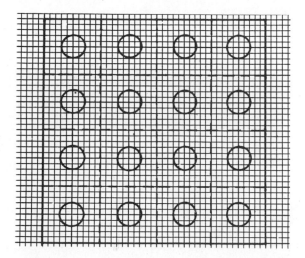

Figure 6–1   WHEN LOCATING LIGHTING OUTLETS IT IS USUALLY DESIRA-
BLE TO ARRANGE THEM IN THE FORM OF SQUARES OR RECTANGLES.

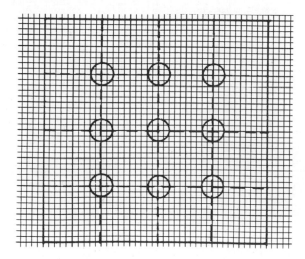

Figure 6–2   WHEN LAYING OUT AN AREA IN SQUARES, THE OUTLETS
SHOULD BE PLACED IN THE CENTERS OF THE SQUARES AND NOT AT THE
CORNERS AS SHOWN HERE.

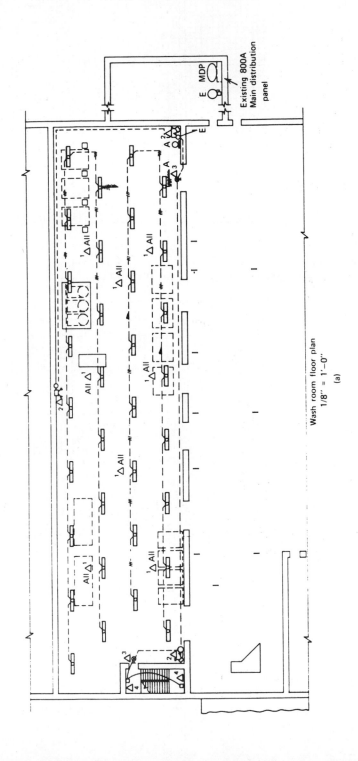

Figure 6-3 LAYOUT OF 8-FT. FLUORESCENT FIXTURES IN AN INDUSTRIAL AREA.

78

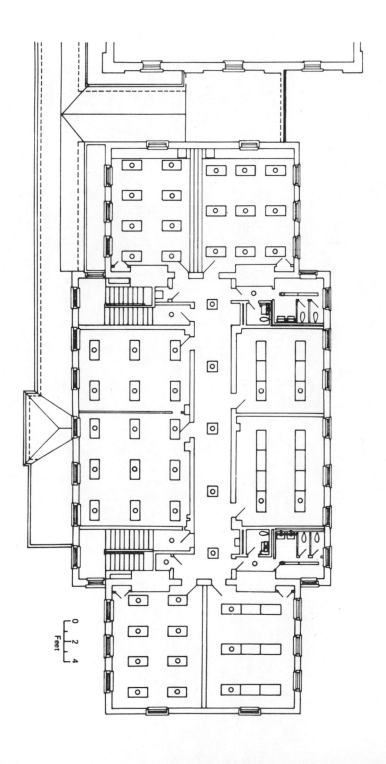

Figure 6-4 LAYOUT OF 2' × 4' AND 2' × 2' FLUORESCENT FIXTURES IN AN INSTITUTIONAL BUILDING.

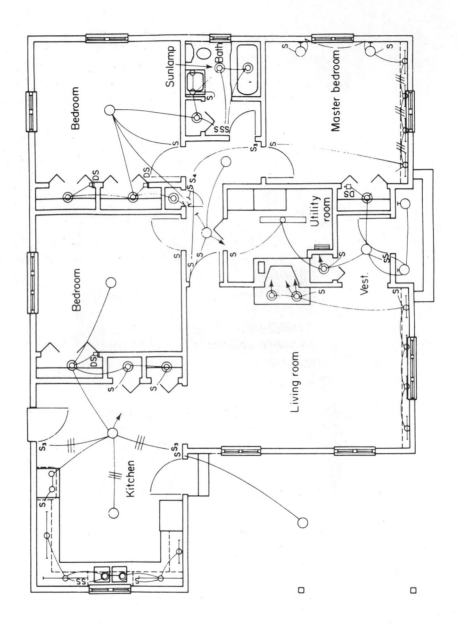

Figure 6–5   LAYOUT OF A RESIDENTIAL LIGHTING DESIGN.

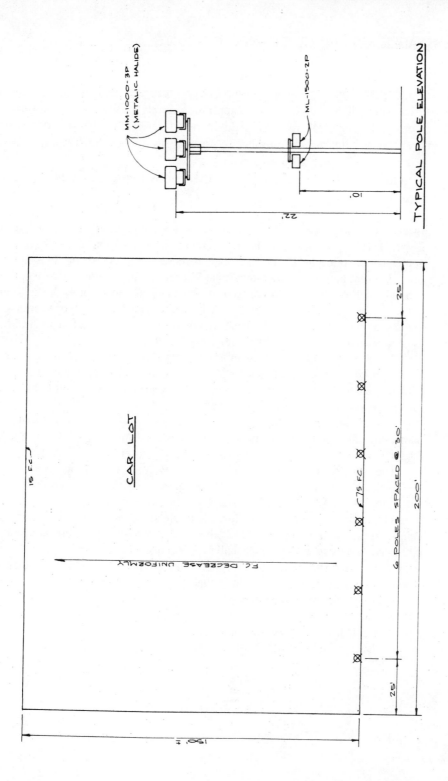

MM-1000-3P
(METALIC HALIDE)

ML-1500-2P

TYPICAL POLE ELEVATION

22'

10'

CAR LOT

15 F.C.

F.C. DECREASE UNIFORMLY

75 FC

25'

25'

6 POLES SPACED @ 30'

200'

150' ±

Figure 6-6   FLOODLIGHT LAYOUT FOR CAR LOT.

6–6   PORTABLE APPLIANCE BRANCH CIRCUITS

In addition to the number of branch circuits described previously, two or more 20-ampere small appliance branch circuits must be provided for all receptacle outlets intended to feed small appliance loads. These loads include refrigerators and freezers in residential kitchens, pantries, breakfast rooms, dining rooms, etc. Regardless of the number of circuits used, they must not have receptacles rated for less than 20 amperes.

Article 220–16 of the NEC has the following comments concerning appliance loads:

> In each dwelling unit the feeder load shall be computed at 1500 watts for each 2-wire small appliance branch circuit required by Section 220–3(b) for small appliances supplied by 15- or 20-ampere receptacles on 20-ampere branch circuits in the kitchen, pantry, dining room, breakfast room, family room. Where the load is subdivided through two or more feeders, the computed load for each shall include not less than 1500 watts for each 2-wire branch circuit for small appliances. These loads shall be permitted for the general lighting load and subjected to the demand factors permitted in Table 220–11 for the general lighting load.
>
> A feeder load of not less than 1500 watts shall be included for each 2-wire laundry branch circuit installed as required by Section 220–3(c). It shall be permissible to include this load with the general lighting load and subjected to the demand factors provided in Section 220–11.
>
> It shall be permissible to apply a demand factor of 75 percent to the nameplate-rating load of four or appliances fastened in place served by the same feeder in a one-family, two-family or multifamily dwelling.
>
> The load for household electric clothes dryers in a dwelling unit(s) shall be 5000 watts or the nameplate rating, whichever is larger, for each dryer served. The use of the demand factors in Table 220–18 shall be permitted.
>
> The feeder demand load for household electric ranges, wall-mounted ovens, counter-mounted cooking units, and other household cooking appliances individually rated in excess of 1-3/4 kW shall be permitted to be computed in accordance with Table 220–19. Where two or more single-phase ranges are supplied by a 3-phase, 4-wire feeder, the total load shall be computed on the basis of twice the maximum number connected between any two phases.
>
> It shall be permissible to compute the load for commercial electric cooking equipment, dishwasher booster heaters, water heaters, and other kitchen equipment in accordance with Table 220–20.

QUESTIONS

Answer the following questions by filling in the blanks.

1   The size of branch circuit conductors varies depending on the _____ requirements of the equipment being fed.

2   The smallest size conductor permitted for use on branch circuits is _____ AWG copper.

3   All receptacles installed on 15- and 20-ampere branch circuits must be of the _____ type.

4   The NEC requires that the total load on a branch circuit must not exceed _____ percent of the circuit rating when the load is in operation for 3 hours or longer.

5   It is good practice to limit the load on 20-ampere branch circuits from between _____ and 1600 watts.

6   In most cases, lighting outlets should be located in a given area so that the illumination is _____.

7   _____ or more 20-ampere small branch circuits must be provided in all residential kitchens.

8   A branch circuit rated at not less than _____ watts must be provided for each 2-wire laundry circuit.

7

PRINCIPLES OF ILLUMINATION

Those required to design lighting systems should consider it from three basic aspects: quantity, quality, and cost. Therefore, every lighting design or layout should provide the highest visual comfort and performance consistent with the type of area to be illuminated and the budget provided by the owners. There will be more than one solution—in most cases—for any lighting application; some are dull and commonplace, whereas others show imagination and resourcefulness on the part of the designer.

In most cases, electrical engineers furnish working drawings and specifications for lighting projects of any consequence. The electrician is only required to interpret these drawings and to install the quantity and types of lighting fixtures specified in the proper locations. However, there are many problems that can arise from working drawings, depending on how well they are prepared, and the electrician must take care to solve them prior to beginning the installation. Solutions to many of these programs are covered in this chapter. This chapter is also designed to give the reader a basic knowledge of illuminating engineering, i.e., how to make lighting calculations for various areas, how to design lighting layouts, and how to select the proper lamp sources and illuminaires (lighting fixtures) for a given application.

7–1   RESIDENTIAL LIGHTING CALCULATIONS

Unlike most commercial and industrial lighting applications, extensive illumination calculations are not required for residential lighting. However, in order to obtain the proper amount of illumination in all areas of the home, some guide should be followed. The possibilities of lighting schemes available are limited only by the electrician's or homeowner's imagination.

The person designing a residential lighting layout should first become thoroughly familiar with the many types of residential lighting fixtures that are currently available. One of the best ways to accomplish this is to obtain several residential lighting catalogs and to study them. With the owner's ideas regarding the types they desire to use in the various areas of the home in mind, the lighting designer may commence work by using a method called the lumens-per-square-foot method to determine the number of lighting fixtures required and the size of lamps (wattage) to use in each.

When using this method, keep in mind that lighter wall and ceiling colors reflect more light and that darker colors absorb more light. The tables contained herein are based on rooms with light ceiling and wall colors; therefore, if the calculations are used for areas with dark-colored surfaces, the total lumens obtained in the calculations should be mutliplied by a factor of at least 1.25 to insure that the proper minimum amount of illumination will be provided. These tables are based on surface-mounted lighting fixtures. If the area to be lighted contains recessed fixtures, the total lumens obtained from the recessed fixtures should be multiplied by a factor of 0.60.

Table7-1   MINIMUM RECOMMENDED ILLUMINATION FOR
VARIOUS RESIDENTIAL AREAS

Area	Lumens required per square foot
Living Room	80
Kitchen	45
Kitchen	80
Bathroom	65
Hallway	45
Laundry	70
Workshop (over workbench)	70
Bedroom	50

Multi-purpose rooms, such as the family—recreation room, should have means of varying the illumination for different activities. This is accomplished either by controlling (switching) several groups of lights or with a dimmer switch.

To begin a lighting calculation for a residential area, the designer must obtain the room dimensions. If scaled drawings are available, room dimensions may be obtained by measuring the dimensions on the drawing with an architect's scale. If drawings are not available, actual measurement of the area with a tape measure is necessary.

A floor plan of a living room in a small residence is shown in Figure 7–1. In scaling this drawing, we will assume that the area 13.75 feet × 19 feet equals 261 square feet. This figure is then mutliplied by the required lumens per square foot, obtained from Table 7–1, to obtain the total lumens required:

$$261 \text{ (ft}^2) \times 80 \text{ (lumens required per ft}^2) = 20{,}880 \text{ lumens.}$$

The next step is to refer to manufacturer's lamp data (see Figure 7–2) to select lamps that will give the required lumens. At the same time, the designer should refer to residential lighting fixture catalogs to get an idea of the types of lighting fixtures to use in the area.

The lighting layout for the living room is shown in Figure 7–1. Here, two recessed "wall-wash" lighting fixtures are mounted in the ceiling in front of the fireplace to highlight the stone facing of the fireplace and chimney. The fixture catalog states that this type of lighting fixture will accept incandescent lamps up to 150 watts. The lamp data in Figure 7–2 indicate that a 150-watt I.F. (inside-frosted) lamp has approximately 2880 initial lumens; thus, the two lamps used in this part of the living room give a total of 5760 lumens. However, since the fixtures are recessed, the total lumen output must be multiplied by a factor of 0.60; this gives a total of only 3456 usable lumens. This means that an additional

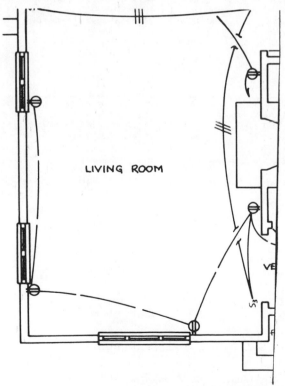

Figure 7–1   FLOOR PLAN OF A LIVING ROOM IN A SMALL RESIDENCE.

60 WATTS

Bulb	Base	Order	Pkg	Volts	Price	Description	Volts	Filament	MOL	LCL	Avg Life	Lumens
A-19	Medium	60A		120	.34	Inside Frosted (11)	120	CC-6	$4\frac{7}{16}$	$3\frac{1}{4}$	1000	870
		60A	24PK PM	120	.34	Inside Frosted. 24-Pack **PRICE MARKED**	120	CC-6	$4\frac{7}{16}$	$3\frac{1}{4}$	1000	870
		60A/TF	24PK	115-125	.76	Inside Frosted—TUFF-SKIN 24-Pack (44)	120	CC-6	$4\frac{7}{16}$	....	1000	...
		60A		125	.37	Inside Frosted (11)	120	CC-6	$4\frac{7}{16}$	$3\frac{1}{4}$	1000	870
		60A		130	.41	Inside Frosted (11)	120	CC-6	$4\frac{7}{16}$	$3\frac{1}{4}$	1000	870
		60A/W	24PK PM	120	.37	Soft-White. 24-Pack **PRICE MARKED** (11)	120	CC-6	$4\frac{7}{16}$	...	1000	855
		60A/WP	24PK PM	120	4/1.98	Soft-White *PLUS*—24-Pack **PRICE MARKED** (11)	120	CC-6	$4\frac{7}{16}$	$3\frac{1}{4}$	1500	820
		60A/CL		120	.39	Clear (11)	120	CC-6	$4\frac{7}{16}$	$3\frac{1}{4}$	1000	870
		60A/CL	24PK PM	120	.39	Clear. 24-Pack **PRICE MARKED** (11)	120	CC-6	$4\frac{7}{16}$	$3\frac{1}{4}$	1000	870
		60A/CL		130	.47	Clear (11)	120	CC-6	$4\frac{7}{16}$	$3\frac{1}{4}$	1000	870

75 WATTS

Bulb	Base	Order	Pkg	Volts	Price	Description	Volts	Filament	MOL	LCL	Avg Life	Lumens
A-19	Medium	75A		120	.34	Inside Frosted (11)	120	CC-6	$4\frac{7}{16}$	$3\frac{1}{4}$	750	1190
		75A	24PK PM	120	.34	Inside Frosted. 24-Pack **PRICE MARKED** (11)	120	CC-6	$4\frac{7}{16}$	$3\frac{1}{4}$	750	1190
		75A		125	.37	Inside Frosted (11)	120	CC-6	$4\frac{7}{16}$	$3\frac{1}{4}$	750	1190
		75A		130	.41	Inside Frosted (11)	120	CC-6	$4\frac{7}{16}$	$3\frac{1}{4}$	750	1190
		75A/W	24PK PM	120	.37	Soft-White. 24-Pack **PRICE MARKED** (11)	120	CC-6	$4\frac{7}{16}$	$3\frac{1}{4}$	750	1170
		75A/WP	24PK PM	120	4/1.98	Soft-White *PLUS*. 24-Pack **PRICE MARKED** (11)	120	CC-6	$4\frac{7}{16}$	$3\frac{1}{4}$	1500	1075
		75A/CL		120	.39	Clear (11)	120	CC-6	$4\frac{7}{16}$	$3\frac{1}{4}$	750	1190
		75A/CL		130	.47	Clear (11)	120	CC-6	$4\frac{7}{16}$	$3\frac{1}{4}$	750	1190
	▲Medium	75A/99		120	.44	Inside Frosted—Extended Service (11)	120	CC-6	$4\frac{7}{16}$	$3\frac{1}{4}$	2500	1000
		75A/99	24PK	120	.44	Inside Frosted—Extended Service. 24-Pack (11)	120	CC-6	$4\frac{7}{16}$	$3\frac{1}{4}$	2500	1000
		75A/99		130	.53	Inside Frosted—Extended Service (11)	120	CC-6	$4\frac{7}{16}$	$3\frac{1}{4}$	2500	1000

100 WATTS

Bulb	Base	Order	Pkg	Volts	Price	Description	Volts	Filament	MOL	LCL	Avg Life	Lumens
A-19	Medium	100A		120	.34	Inside Frosted—Bonus Line (46)	120	CC-8	$4\frac{7}{16}$	$3\frac{1}{4}$	750	1750
		100A	24PK PM	120	.34	Inside Frosted—Bonus Line. 24-Pack **PRICE MARKED** (46)	120	CC-8	$4\frac{7}{16}$	$3\frac{1}{4}$	750	1750
		100A		125	.37	Inside Frosted—Bonus Line (46)	120	CC-8	$4\frac{7}{16}$	$3\frac{1}{4}$	750	1750
		100A		130	.41	Inside Frosted—Bonus Line (46)	120	CC-8	$4\frac{7}{16}$	$3\frac{1}{4}$	750	1750
		100A/W	24PK PM	120	.37	Soft-White—Bonus Line. 24-Pack **PRICE MARKED** (46)	120	CC-8	$4\frac{7}{16}$	..	750	1710
		100A/WP	24PK PM	120	4/1.98	Soft-White *PLUS*. 24-Pack **PRICE MARKED** (11)	120	CC-8	$4\frac{7}{16}$	...	1500	1585
		100A/CL		120	.39	Clear—Bonus Line (46)	120	CC-8	$4\frac{7}{16}$	$3\frac{1}{4}$	750	1750
		100A/CL	24PK PM	120	.39	Clear—Bonus Line. 24-Pack **PRICE MARKED** (46)	120	CC-8	$4\frac{7}{16}$	$3\frac{1}{4}$	750	1750
		100A/CL		130	47	Clear—Bonus Line (46)	120	CC-8	$4\frac{7}{16}$	$3\frac{1}{4}$	750	1750
	▲Medium	100A/99		120	47	Inside Frosted—Extended Service (46)	120	CC-8	$4\frac{7}{16}$	$3\frac{1}{4}$	2500	1490
		100A/99	24PK	120	47	Inside Frosted—Extended Service. 24-Pack (46)	120	CC-8	$4\frac{7}{16}$	$3\frac{1}{4}$	2500	1490
		100A/99		130	57	Inside Frosted—Extended Service (46)	120	CC-8	$4\frac{7}{16}$	$3\frac{1}{4}$	2500	1490
	▲Left-Hand Medium	100A/LHT		120	41	Inside Frosted—Left-hand threaded base (11)	120	CC-8	$4\frac{7}{16}$	$3\frac{1}{4}$	750	1750
		100A/LHT		130	49	Inside Frosted—Left-hand threaded base (11)	120	CC-8	$4\frac{7}{16}$	$3\frac{1}{4}$	750	1750

Figure 7–2 TABLE OF MANUFACTURERS' LAMP DATA.

100-200-300 WATTS

PS-25 	3-Contact Mogul	100/300	6PK PM	120	$1.39	Soft-White—3-Way. Burn base down. 6-Pack **PRICE MARKED**	24	C-2R CC-8	6¹¹⁄₁₆	. . .	1500, 1200, 1150	1320, 3620, 4940
		100/300/DPK	12PK PM	115-125	1.69	Coloramic—Dawn Pink. 3-Way. Burn base down. 12-Pack **PRICE MARKED**	60	C-2R CC-8	6¹¹⁄₁₆	. . .	1200	. . .
		100/300/2		120	2.05	Soft-White—3-Way. Neck of bulb coated red from base to maximum bulb diameter. Burn base down	60	C-2R CC-8	6¹¹⁄₁₆	. . .	1200	. . .

Nominal Lamp Watts	Bulb	Nominal Length (Inches)	Base	Lamp Ordering Code	List Price	DESCRIPTION (See Fluorescent Lamp Footnotes—Page 63)	Std. Pkg. Qty.	Approx. Hours Life	Approx. Initial Lumens	Approx. Lumens at 40% Rated Avg.Life

T-12   Approx. 1½″ Diameter
To determine approximate lamp length (pin ends to pin ends), deduct 1/4″ from the nominal lamp length shown below.

40	T-12	48	Medium Bipin 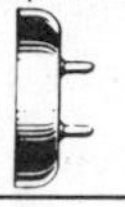	F40CW/S		$1.49	Cool White—STAYBRIGHT	24	15000†	3250	2960
				F40CW/S	6PK	1.67	Cool White—STAYBRIGHT 6-Pack	24	15000†	3250	2960
				F40D/S		1.70	Daylight—STAYBRIGHT	24	15000†	2650	2410
				F40W/S		1.70	White—STAYBRIGHT	24	15000†	3300	3005
				F40WW/S		1.70	Warm White—STAYBRIGHT	24	15000†	3300	3005

Figure 7–2   (Concluded)

17,424 will be required to reach the recommended values; that is, 20,880 (total lumens required in calcuation) − 3456 (usable lumens obtained from the two recessed fixtures) equals 17,424 (lumens to account for).

The owner's preference for the living room in question was for some type of indirect lighting. Therefore, a drapery cornice is placed along the entire front wall of the area, with four 40-watt single-tube fluorescent fixtures installed behind the cornice. Each warm-white fluorescent lamp is rated at 2080 lumens, which gives a total of 8320 lumens for the four fixtures. When this figure is added to the lumens of the recessed wall-wash fixtures, only 9104 lumens must yet be accounted for.

Although not shown on the floor plan in Figure 7–1, two three-way lamps (100, 200, and 300 watts) are used in table lamps on end tables located on each end of a sofa. This gives an additional 9460 lumens in the area, for a total of 21,236 lumens—close enough to the total recommendations to be nearly perfect. As a final touch, dimmers are added to control the recessed fixtures at the fireplace and the cornice fluorescent lights. Since the two three-way lamps can be dimmed by switching to different wattages, the lighting levels in this area can be varied to suit the activities: low for a relaxed method or bright for a gay party mood.

This method makes it possible to determine quickly and accurately the number and size of light sources needed to achieve the recommended lighting level in any area of the home.

Selecting the Lighting Fixtures

Since there are numerous variables involved in selecting the types of lighting fixtures for a given application, no set rules are available. However, the following guidelines should prove useful:

1   Determine the total lumens required for a given area from Table 7–1 and by other methods.

2   Study residential lighting catalogs to see what types of lighting fixtures are available. Also study interior decorating magazines for ideas.

3   Prepare a master plan of the area. Then select a fixture or fixtures that will fit into the architectural or decorating scheme of the area to be lighted.

4   Read the manufacturer's description in the catalog when selecting fixtures to find out the number and size of lamps recommended for use in those selected. Then look at the data (Figure 7–2) given for the lamps in order to obtain the lumen output of the lamp(s).

Additional hints on selecting proper lighting for various areas in the home are presented below. The separate areas are broken down so that each may be discussed individually.

Living room   This is the area in the home where guests are entertained and where the family gathers to relax, watch TV, or engage in conversation. Lighting in this area should emphasize any special architectural features, such as planters or bookcases. Pull-down lighting fixtures or table lamps placed near chairs or on end tables are used to supply reading light.

Dining and kitchen areas   In residential dining rooms, a chandelier mounted directly above the dining table and controlled by a dimmer/switch becomes the centerpiece of the room while providing general illumination. The dimmer adds versatility since the lighting can be dimmed for formal dining or made bright for an evening of cards.

In addition to the center chandelier, supplementary lighting at the buffet and sideboards is often desirable. Use recessed accent lights for a contemporary design and wall brackets to match the chandelier for a traditional setting. Other possibilities include using concealed fluorescent lighting in valances or cornices.

The ideal general lighting system for a residential kitchen would be a luminous ceiling. This type of lighting arrangement gives a skylight effect but is also the most expensive to install. The effect is achieved by installing rows of bare fluorescent strip-lighting fixtures above a dropped ceiling consisting of ceiling panels with attractive diffuser patterns. The fixtures should be spaced approximately 2 feet apart on centers to obtain the recommended illumination level.

If the luminous ceiling is not used, a fixture mounted in the center of the kitchen area will provide general illumination. Additional lights should be mounted over the sink, electric range, and under wall cabinets to provide light on the work surfaces.

Bedrooms Bedroom lighting should be both decorative and functional, with flexibility of control in order to create the desired lighting environment. For example, reading and sewing are common activities occurring in the bedroom, and both require good illumination to lessen eye strain. Other activities, however, such as casual conversation or watching TV, require only general, non-glaring room illumination, preferably controlled by a dimmer switch. Proper lighting in and around the closet area can do much to help in the selection and appearance of clothes, and supplementary lighting around the vanity will aid in personal grooming.

Bathrooms Good light is needed in all bathrooms for good grooming and hygiene practices. If the bath is small, the mirror light combined with a tub or shower light will usually suffice. On the other hand, if the bathroom is large, a bright, central light source is recommended, and supplemental light at the mirror should be provided. Luminous ceilings are also becoming popular in the bathroom.

Basement, utility room, and workshop All these areas require a similar amount of illumination and lighting techniques. The general lighting need only be about 45 lumens per square foot, but supplemental light over work areas of at least 70 lumens per square foot (of work surface) should be provided. Since these areas will normally not be visited by guests, inexpensive lighting fixtures can be used.

Family room A well designed lighting layout for the family room includes graceful blending of general lighting to illuminate the overall area with well chosen supplemental lighting to aid certain individual seeing tasks. For example, diffused, recessed, incandescent, lighting fixtures installed flush with the ceiling of the family room will furnish even, glare-free light throughout the room—if the proper numbers of fixtures are installed and spaced correctly.

Lamps concealed behind cornices near the ceiling will enrich the natural beauty of paneled walls or the texture of brick, natural stone walls, and the like. This technique is also very effective over bookshelves where the light is positioned to shine on books with colorful bindings. Fluorescent lamps concealed in a cove lighting system will not only furnish excellent, indirect, general illumination for a family room, but will also give the impression of a higher ceiling; this is a very desirable effect in low-ceiling family rooms in the basement of homes.

Keep in mind that the lighting layout for any family room should be highly flexible since this is an area that will be used for a variety of daily activities. To illustrate, casual conversation is enhanced amid subdued, complexion-flattering light, such as incandescent or warm-white, fluorescent lamps controlled by a dimmer switch; game participants feel more comfortable in a uniformly lighted room with some additional glare-free light directed onto the playing areas; low-level lighting over the bar area should be just bright enough for mixing a drink or having a snack. TV viewing requires only softly lighted surroundings whereas reading requires a somewhat brighter light source, with light directly on the pages.

Example 7–1

The floor plan in Figure 7–3 shows the second floor of a modified A-frame cabin. Use the data given in this chapter to find:

a   The required light flux in lumens for bedroom A.

b   The required light flux for bedroom B.

c   The required light flux for the bathroom.

Solution

a   13.75 feet $\times$ 14 feet (by scaling drawings) = 192.5 (ft²) $\times$ 50 lumen/ft² (from Table 7–1) = 9625 lumens.

b   14.75 feet $\times$ 14 feet = 206.5 ft² $\times$ 50 lumen/ft² = 10,325 lumens.

c   5.33 feet $\times$ 6.75 feet = 65 ft² $\times$ 65 lumen/ft² = 2338 lumens.

Example 7–2

Use the lamp data in Figure 7–2 to find the required number of lamps for the rooms in Example 7–1 if:

a   Bedroom A will be lighted with 40-watt fluorescent lamps (F40WW/S).

b   Bedroom B will be illuminated with 100-watt incandescent lamps (100 watt inside frosted—bonus line).

c   Bathroom will be lighted with 60-watt I.F. incandescent lamps.

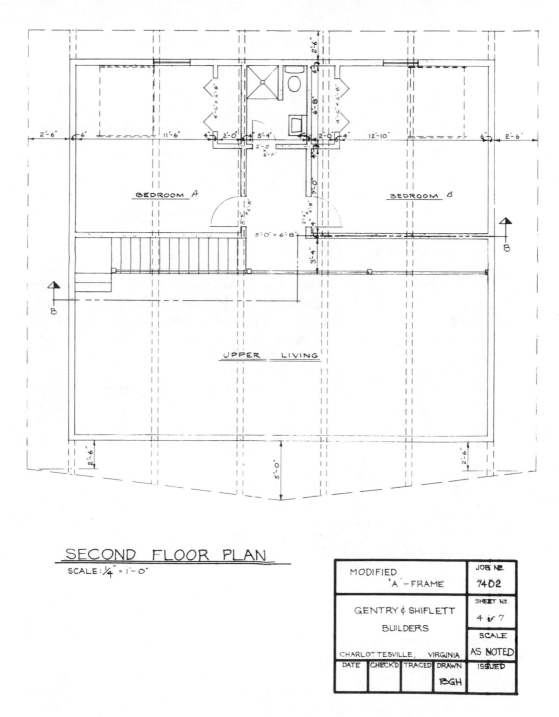

SECOND FLOOR PLAN

SCALE: ¼" = 1'-0"

MODIFIED 'A'-FRAME	JOB № 7402
GENTRY & SHIFLETT BUILDERS	SHEET № 4 of 7
	SCALE AS NOTED
CHARLOTTESVILLE, VIRGINIA	
DATE	CHECK'D	TRACED	DRAWN BGH	ISSUED

Figure 7–3 FLOOR PLAN SHOWING THE SECOND FLOOR LEVEL OF A MODIFIED A-FRAME CABIN.

Solution

a   Bedroom A requires 9625 lumens, and a 40-watt F40WW/S lamp has approximately 3300 initial lumens. Therefore $\dfrac{9625 \text{ lumens}}{3300 \text{ lumens/lamp}}$ 2.91 or 3 lamps.

b   Bedroom B requires 10,325 lumens, and a 100-watt I.F. lamp has approximately 1750 initial lumens. Thus, $\dfrac{10,325 \text{ lumens}}{1750 \text{ lumens/lamp}} = 5.9$ or 6 lamps.

c   The bathroom requires 2338 lumens, and a 60-watt I.F. incandescent lamp has approximately 870 initial lumens. Therefore, $\dfrac{2338 \text{ lumens}}{870 \text{ lumens/lamp}} = 2.68$ or 3 lamps.

7–2   ZONAL-CAVITY METHOD

The Illuminating Engineering Society (IES) zonal-cavity method of lighting calculations is used to determine the average maintained-illumination level on the work plane in a given lighting installation and also to determine the number of lighting fixtures (of a particular type) required in a given area to provide the desired or recommended illumination level.

The illumination calculation sheet shown in Figure 7–4 is recommended when the zonal-cavity method is used. The illustration at the bottom of this form shows that a room (area) is separated into three ratios: ceiling cavity ratio (CCR), room cavity ratio (RCR) and floor cavity ratio (FCR). The cavity ratios of the three areas are found as follows:

$$\text{Room-Cavity Ratio} = \frac{5hrc\ (L + W)}{L \times W} \tag{7–1}$$

$$\text{Ceiling-Cavity Ratio} = \frac{5hcc\ (L + W)}{L \times W} \tag{7–2}$$

$$\text{Floor-Cavity Ratio} = \frac{5hfc\ (L + W)}{L \times W} \tag{7–3}$$

where hrc, hcc, and hfc are the cavity heights as shown in Figure 7–4, L is the room length and W is the room width. From Equations 7–1, 7–2, and 7–3, we see that the cavity ratio for any area is found by mutliplying the height of the cavity in question by five times the sum of the area's length and width and then dividing by the product of the area's length and width.

To begin a calculation for an area using the IES zonal-cavity method, record the room width and length, the ceiling height, and the mounting height of the lighting fixtures above the floor and fill in all other manufacturers' data shown in the form in Figure 7–4. Once all of the required information is inserted in the proper places on the form, calculate the three cavity ratios using Equations 7–1, 7–2, and 7–3. Insert the resulting data in the spaces provided for RCR, CRR, and FCR.

ILLUMINATION CALCULATION SHEET

For Use with the IES Zonal Cavity Method

General Information

Project identification: _____

Average Maintained illumination for Design: _____ footcandles.

Luminaire Data:

Manufacturer: _____ Catalog Number: _____

Lamps (type & color): _____ Number per Luminaire: _____

Total Lumens per Luminaire: _____ Maintenance Factor: _____

Selection of Coefficient of Utilization

Step 1: Fill in sketch at right.

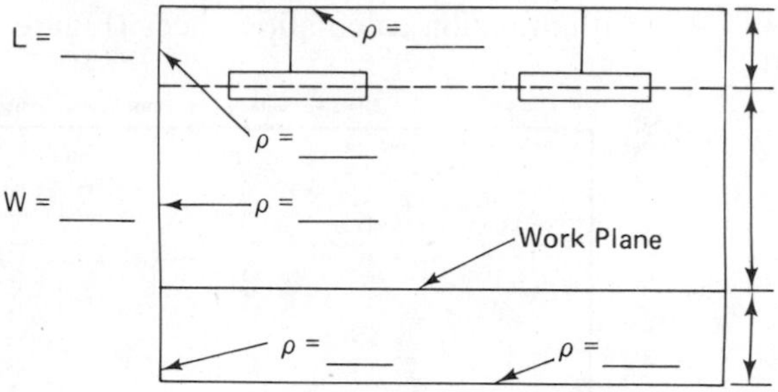

Step 2: Determine cavity ratios by the formula:

$$\frac{5 \times h \begin{bmatrix} CC \\ RC \\ FC \end{bmatrix} \times (L + W)}{L \times W}$$

Room cavity ratio, RCR = _____

Ceiling cavity ratio, CCR = _____

Floor cavity ratio, FCE = _____

L = _____

W = _____

ρ = _____

ρ = _____

ρ = _____

ρ = _____

ρ = _____

Work Plane

Step 3: Obtain effective ceiling cavity reflectance (ρCC).   ρCC = _____

Step 4: Obtain effective floor cavity reflectance (ρFC).   ρFC = _____

Step 5: Obtain coefficient of utilization (CU) from manufacturer's data.   CU = _____

Calculations

Step 6:

Average Maintained Illumination Level

$$\text{Footcandles} = \frac{(\text{Total lamp lumens per Luminaire}) \times (CU) \times \text{maintenance factor}}{\text{area per Luminaire}}$$

= _____

= _____ footcandles on work area

Area per Luminaire: (This area divided by the Luminaire length gives the approximate spacing between continuous row, or it may be divided into the total room area to determine the number of Luminaires required.)

$$\text{Area per Luminaire} = \frac{(\text{Total lamp lumens per Luminaire}) \times (CU) \times \text{maintenance factor}}{\text{footcandles}}$$

= _____

= _____ Square feet

Calculated by: _____ Date: _____

Figure 7–4   CALCULATION FORM FOR ZONAL CAVITY METHOD.

The next step is to select the effective ceiling reflectance (Pcc) from Figure 7–5 for the actual combination of ceiling and wall reflectances. Note that for surface-mounted or recessed lighting fixtures, CCR will equal 0, and the ceiling reflectance may be used as the effective cavity reflectance. Continue by selecting the effective floor-cavity reflectance (Pfc) for the combination on floor and wall reflectances, also from Figure 7–5. Enter both these reflectance values in the illumination calculation sheet (Figure 7–4).

The coefficient of utilization (cu) is determined by referring to a "coefficient of utilization table" for the lighting fixture under consideration (see Figure 7–6). Coefficient of utilization tables are normally supplied in manufacturer's catalogs for each type of lighting fixture; If they are not shown in the catalog, write the manufacturer for photometric data for the fixture in question. The CU is a measure of the total light flux received by a surface divided by the total flux from the lamps illuminating it. When the CU has been determined, it should be entered in the illumination calculation sheet (Figure 7–4).

Effective Ceiling- or Floor-Cavity Reflectance for Various Reflectance Combinations

PER CENT CEILING OR FLOOR REFLECTANCE	90				80				70			50			30				10		
PER CENT WALL REFLECTANCE / Ceiling or Floor Cavity Ratio	90	70	50	30	80	70	50	30	70	50	30	70	50	30	65	50	30	10	50	30	10
0	90	90	90	90	80	80	80	80	70	70	70	50	50	50	30	30	30	30	10	10	10
0.1	90	89	88	87	79	79	78	78	69	69	68	59	49	48	30	30	29	29	10	10	10
0.2	89	88	86	85	79	78	77	76	68	67	66	49	48	47	30	29	29	28	10	10	9
0.3	89	87	85	83	78	77	75	74	68	66	64	49	47	46	30	29	28	27	10	10	9
0.4	88	86	83	81	78	76	74	72	67	65	63	48	46	45	30	29	27	26	11	10	9
0.5	88	85	81	78	77	75	73	70	66	64	61	48	46	44	29	28	27	25	11	10	9
0.6	88	84	80	76	77	75	71	68	65	62	59	47	45	43	29	28	26	25	11	10	9
0.7	88	83	78	74	76	74	70	66	65	61	58	47	44	42	29	28	26	24	11	10	8
0.8	87	82	77	73	75	73	69	65	64	60	56	47	43	41	29	27	25	23	11	10	8
0.9	87	81	76	71	75	72	68	63	63	59	55	46	43	40	29	27	25	22	11	9	8
1.0	86	80	74	69	74	71	66	61	63	58	53	46	42	39	29	27	24	22	11	9	8
1.1	86	79	73	67	74	71	65	60	62	57	52	46	41	38	29	26	24	21	11	9	8
1.2	86	78	72	65	73	70	64	58	61	56	50	45	41	37	29	26	23	20	12	9	7
1.3	85	78	70	64	73	69	63	57	61	55	49	45	40	36	29	26	23	20	12	9	7
1.4	85	77	69	62	72	68	62	55	60	54	48	45	40	35	28	26	22	19	12	9	7
1.5	85	76	68	61	72	68	61	54	59	53	47	44	39	34	28	25	22	18	12	9	7
1.6	85	75	66	59	71	67	60	53	59	52	45	44	39	33	28	25	21	18	12	9	7
1.7	84	74	65	58	71	66	59	52	58	51	44	44	38	32	28	25	21	17	12	9	7
1.8	84	73	64	56	70	65	58	50	57	50	43	43	37	32	28	25	21	17	12	9	6
1.9	84	73	63	55	70	65	57	49	57	49	42	43	37	31	28	24	20	16	12	9	6
2.0	83	72	62	53	69	64	56	48	56	48	41	43	37	30	28	24	20	16	12	9	6
2.1	83	71	61	52	69	63	55	47	56	47	40	43	36	29	28	24	20	16	13	9	6
2.2	83	70	60	51	68	63	54	45	55	46	39	42	36	29	28	24	19	15	13	9	6
2.3	83	69	59	50	68	62	53	44	54	46	38	42	35	28	28	24	19	15	13	9	6
2.4	82	68	58	48	67	61	52	43	54	45	37	42	35	27	28	24	19	14	13	9	6
2.5	82	68	57	47	67	61	51	42	53	44	36	41	34	27	27	23	18	14	13	9	6
2.6	82	67	56	46	66	60	50	41	53	43	35	41	34	26	27	23	18	13	13	9	5
2.7	82	66	55	45	66	60	49	40	52	43	34	41	33	26	27	23	18	13	13	9	5
2.8	81	66	54	44	66	59	48	39	52	42	33	41	33	25	27	23	18	13	13	9	5
2.9	81	65	53	43	65	58	48	38	51	41	33	40	33	25	27	23	17	12	13	9	5
3.0	81	64	52	42	65	58	47	38	51	40	32	40	32	24	27	22	17	12	13	8	5
3.1	80	64	51	41	64	57	46	37	50	40	31	40	32	24	27	22	17	12	13	8	5
3.2	80	63	50	40	64	57	45	36	50	39	30	40	31	23	27	22	16	11	13	8	5
3.3	80	62	49	39	64	56	44	35	49	39	30	39	31	23	27	22	16	11	13	8	5
3.4	80	62	48	38	63	56	44	34	49	38	29	39	30	22	26	22	16	11	13	8	5
3.5	79	61	48	37	63	55	43	33	48	38	29	39	30	22	26	22	16	11	13	8	5
3.6	79	60	47	36	62	54	42	33	48	37	28	39	30	21	26	21	15	10	13	8	5
3.7	79	60	46	35	62	54	42	32	48	37	27	38	30	21	26	21	15	10	13	8	4
3.8	79	59	45	35	62	53	41	31	47	36	27	38	29	21	26	21	15	10	13	8	4
3.9	78	59	45	34	61	53	40	30	47	36	26	38	29	20	26	21	15	10	13	8	4
4.0	78	58	44	33	61	52	40	30	46	35	26	38	29	20	26	21	15	9	13	8	4
4.1	78	57	43	32	60	52	39	29	46	35	25	37	28	20	26	21	14	9	13	8	4
4.2	78	57	43	32	60	51	39	29	46	34	25	37	28	19	26	20	14	9	13	8	4
4.3	78	56	42	31	60	51	38	28	45	34	25	37	28	19	26	20	14	9	13	8	4
4.4	77	56	41	30	59	51	38	28	45	34	24	37	27	19	26	20	14	8	13	8	4
4.5	77	55	41	30	59	50	37	27	45	33	24	37	27	18	25	20	14	8	14	8	4
4.6	77	55	40	29	59	50	37	26	44	33	24	36	27	18	25	20	14	8	14	8	4
4.7	77	54	40	29	58	49	36	26	44	33	23	36	26	18	25	20	13	8	14	8	4
4.8	76	54	39	28	58	49	36	25	44	32	23	36	26	18	25	19	13	8	14	8	4
4.9	76	53	38	28	58	49	35	25	44	32	23	36	26	17	25	19	13	7	14	8	4
5.0	76	53	38	27	57	48	35	25	43	32	22	36	26	17	25	19	13	7	14	8	4

Figure 7–5   TABLE OF CEILING, WALL, AND FLOOR REFLECTANCE.

Coefficients of Utilization
Zonal Cavity Method

Effective Floor Cavity Reflectance—20% ρfc									
Effective Ceiling Cavity Reflectance ρcc		80%			50%			10%	
% Wall Reflectance ρw		50%	30%	10%	50%	30%	10%	50%	30%	10%
	1	.75	.70	.67	.67	.64	.61	.57	.55	.53
	2	.64	.57	.52	.57	.52	.48	.49	.46	.43
	3	.55	.48	.42	.49	.44	.39	.42	.39	.35
	4	.49	.41	.36	.44	.38	.33	.38	.34	.30
	5	.42	.35	.29	.38	.32	.28	.33	.28	.25
	6	.38	.30	.25	.34	.28	.23	.29	.25	.21
	7	.34	.28	.22	.31	.25	.20	.27	.22	.19
	8	.30	.23	.18	.27	.21	.17	.24	.19	.16
	9	.27	.20	.16	.24	.19	.15	.21	.17	.14
	10	.25	.18	.14	.22	.17	.13	.20	.15	.12

Left side annotations:
NADIR C.P.-560
3100 Lumen Lamps

MAINTENANCE FACTORS
Good .75   Med. .70
        Poor .65

Average Brightness in the
60°-90° zone from nadir
shall not exceed 600 Foot-
lamberts endwise or 1150
Footlamberis crosswise.

(vertical label) Room Cavity Ratios

Maximum Spacing to Mounting Height Ratio Above Work Plane is: 1.36

Figure 7–6 TYPICAL TABLE GIVING COEFFICIENT OF UTILIZATION DATA FOR A PARTICULAR TYPE OF LIGHTING FIXTURE.

The maintenance factor (MF) is an estimation determined by several factors: the amount of dirt accumulation on the fixture prior to cleaning; the frequency of cleaning; the aging of the lamps; and the frequency of lamp replacement. This figure can vary, and it takes some experience to select the right one. However, Table 7–2 will serve as a guide:

Table 7–2 MAINTENANCE FACTOR TABLE

Environment	MF
Very clean surroundings, such as hospitals	0.80
Clean surroundings, such as restaurants	0.75
Average surroundings, such as offices and schools	0.70
Below-average surroundings	0.65
Dirty surroundings	0.55

The manufacturer's catalog number is entered in the illumination sheet space marked "Lamp Type," the number of lamps used in the fixture is entered next, and then the total watts per fixture. Lamp lumens can be found in lamp manufacturers' catalogs under the type of lamp used in the fixture. Remaining calculations should be obvious by referring to step 6 in the form shown in Figure 7–4.

Example 7–3

For the room shown in Figure 7–7, the lighting fixtures are suspended 1 foot below the ceiling. The ceiling height is 8.5 feet and the desk top (work plane) is 2.5 feet above the finished floor. Calculate:

a   The ceiling cavity ratio, CCR.

b   The floor cavity ratio, FCR.

c   The room cavity ratio, RCR.

Solution

a $\text{CCR} = 5 \times 1 \quad \dfrac{(12 + 8)}{12 \times 8} = 1.04 \text{ or } 1$

b $\text{FCR} = 5 \times 2.5 \quad \dfrac{(12 + 8)}{12 \times 8} = 2.6 \text{ or } 3$

c   The height of the room cavity is determined by the total height of the room (8.5) minus the ceiling cavity (1 foot) and the floor cavity (2.5 feet).

$$\text{RCR} = 5 \times 5 \quad \dfrac{(12 + 8)}{12 \times 8} = 5.2 \text{ or } 5$$

Example 7–4

Assume that the room shown in Figure 7–7 has a ceiling reflectance of 80 percent and a wall reflectance of 50 percent. Find the effective reflectance.

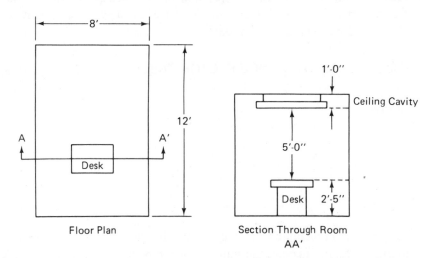

Figure 7–7   FLOOR PLAN IN SECTION OF ROOM USED IN EXAMPLE 7–3.

Solution

Locate the column in Figure 7–5 that contains the known percentage of ceiling (or floor) reflectance (80 percent) and wall reflectance (50 percent). Then move down the column to the line closest to the calculated ceiling-cavity ratio (1.0). The intersection of these horizontal and vertical values indicates that the effective ceiling reflectance is 66 percent.

7–3   SELECTING LIGHTING FIXTURES

The selection of lighting fixtures for any installation involves the consideration of many variables. What is the purpose of lighting the installation—is it for critical seeing, selling, or decoration? How severe is the seeing task, and for what length of time is it to be performed? What are the architectural and decorative requirements, together with the constructional limitations, of the area? What economic considerations are involved? The answers to questions such as these determine the amount of light that should be provided, and the best means of providing it. Since individual tastes and opinions vary, especially in matters of appearance, no one solution of a lighting problem is the most desirable under all circumstances. However, certain basic rules governing adequate quantity and good quality should always be observed when lighting fixtures are being selected.

Proper candlepower distribution for the particular lighting application should be the first consideration in selection of lighting fixtures. Fixtures should be chosen for distribution characteristics suitable to the requirements of the given situation.

The efficiency of a lighting fixture is one measure of the quality of its material and design. Any control (diffuser, grids, etc.) applied to the light output of a bare lamp results in some absorption of light. Usually, the greater the degree of control, the greater the light loss and the lower the efficiency. In many installations the use of low-efficiency lighting fixtures is justifiable in order to achieve the desired effect. Therefore, it is impractical to compare the efficiencies of dissimilar types of fixtures. However, lighting fixtures that produce the same type of control can be compared on an efficiency basis, and those with higher efficiencies are preferred.

The appearance of the lighting fixture should be given careful consideration with respect to the architecture and decoration of the area in which it is to be used. The requirements depend to some extent on whether the fixtures are functional, decorative, or both. In any event, they should harmonize with the surroundings in architectural style, size, and decorative motif.

7–4   LOCATING LIGHTING FIXTURES

There is no general rule for locating lighting fixtures for general illumination, but it is desirable to place lighting fixtures so that the illumination of a given area is uniform. Supplemental lighting can be used to highlight certain architectural features or to provide higher illumination levels at certain points to facilitate seeing.

The floor plan in Figure 7–8 shows the lighting layout for the renovation of an existing school building. Note that 2 foot × 4 foot fluorescent lay-in fixtures were used in most areas. They are located in such a manner that the distance between fixtures is approximately twice that of the end fixtures to the wall because the end fixtures must provide a certain amount of illumination from the fixture to the wall with no additional help whereas the area between two fixtures receives light from both light sources. In one section of the corridor, 2 foot × 2 foot fixtures with fluorescent ''U'' lamps are used because of structural reasons that prevent the use of 2 foot × 4 foot fixtures in this small area. The lavatories also use fluorescent lamps, but the fixtures are narrow, single-tube types, and the fixtures over the lavatories are wall mounted.

A lighting layout for an industrial building is shown on the drawing in Figure 7–9. The corresponding ''Lighting Fixture Schedule'' (Figure 7–10) gives data pertaining to the fixtures used. The Type 2 fixtures are eight feet in length, and each contains two 110-watt fluorescent lamps. The end rows of fixtures are mounted approximately 4 feet from the walls, while the other rows of fixtures are mounted 10 feet apart, on centers. The fixtures are to be mounted on Kindorf channel, a detail of which appears in Figure 7–11. Notes, lines, and symbols appearing on the drawing indicate the number of circuits, homeruns, etc.

7–5   AREA LIGHTING

Many factors have caused outdoor lighting use to increase at a phenomenal rate recently. Automobile use has increased the need for additional street and highway lighting. Other uses are electrical advertising, luminous decorations, flood-lighting, decorative lighting, and lighting facilities for outdoor sports and recreational areas.

The principal difference between interior and exterior lighting lies in the types of light sources used. Interior lighting can usually be located in the ceiling directly above the area to be lighted. This illumination of interior spaces is aided by the reflection properties of the ceilings, walls, and floors. In outdoor lighting applications, the light sources in many cases must be located outside and a further distance away from the area to be illuminated. Because of this, outdoor light sources must be of comparatively higher wattage and designed to control and direct the light to the desired area accurately.

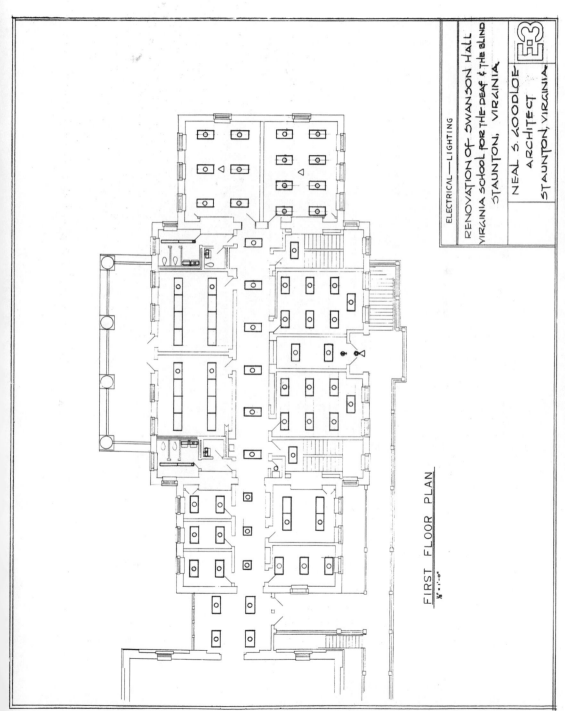

FIRST FLOOR PLAN
⅛"=1'-0"

Figure 7-8   FLOOR PLAN SHOWING THE LIGHTING LAYOUT FOR A RENOVATED SCHOOL BUILDING.

101

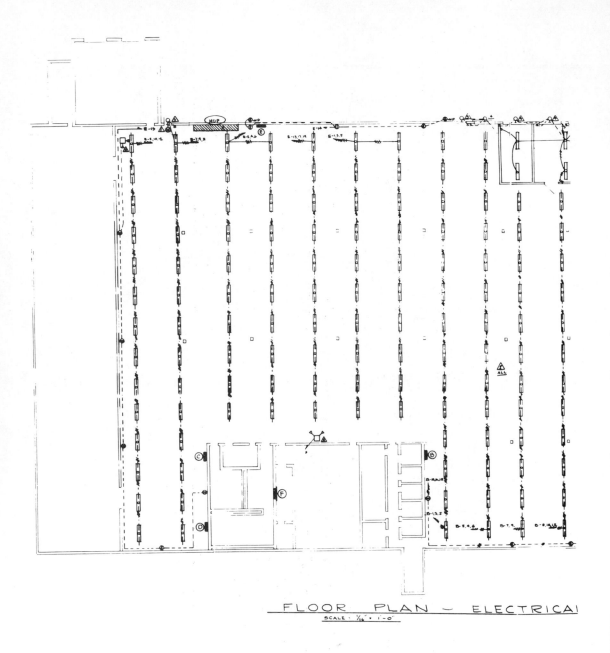

FLOOR PLAN ~ ELECTRICAL
SCALE: 1/16" = 1'-0"

Figure 7–9 LIGHTING LAYOUT FOR AN INDUSTRIAL BUILDING.

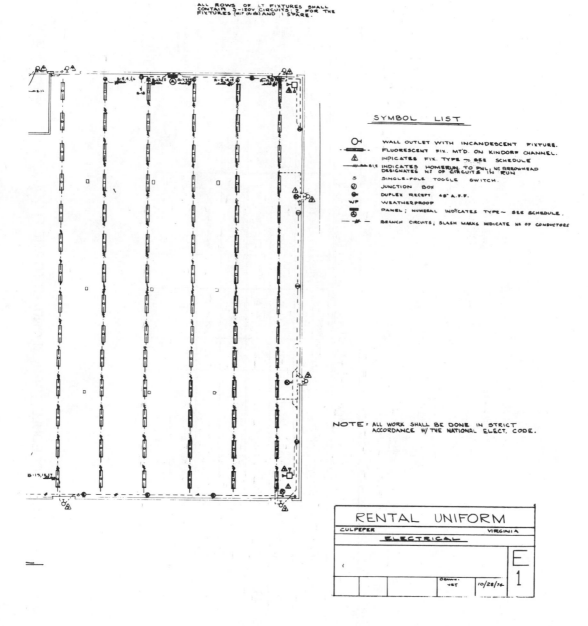

Figure 7–9   (Continued)

LIGHTING FIXTURE SCHEDULE

FIXT TYPE	MANUFACTURER'S DESCRIPTION	LAMPS		VOLTS	MOUNTING	REMARKS
		Nº	TYPE			
1	LITHONIA CAT. Nº EP1R	2	25W I	120	WALL	FEED ALL W/ Nº10 WIRE & CONNECT TO EMERG. SWITCH
2	LITHONIA CAT. Nº AF 2#6 HO PO	2	110W F	120	ON KINDORF	W/ COOL WHITE LAMPS
3	HALO CAT. Nº H2401	1	150W I	120	WALL	MT'D AS DIRECTED BY OWNER
4	HALO CAT. Nº H2411-2	2	100W I	120	WALL	" " "
5	TWIN BEAM BATTERY PACKS					FURNISHED BY OWNERS INSTALLED BY CONTRACTOR
6						

Figure 7-10 LIGHTING FIXTURE SCHEDULE CORRESPONDING TO THE FLOOR PLAN IN FIGURE 7-9.

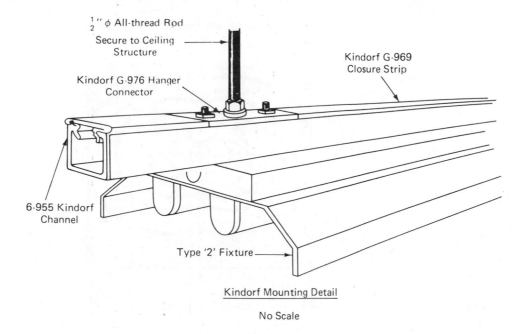

$\frac{1}{2}''$ φ All-thread Rod

Secure to Ceiling
Structure

Kindorf G-969
Closure Strip

Kindorf G-976 Hanger
Connector

6-955 Kindorf
Channel

Type '2' Fixture

Kindorf Mounting Detail

No Scale

Figure 7–11   DETAIL OF A "TYPE 2" LIGHTING FIXTURE MOUNTED ON KIN-DORF CHANNELS.

Good judgment plays an important role in selecting outdoor lighting equipment. In general, however, outdoor lighting is simply a matter of delivering a sufficient (and efficient) quantity of light (measured in lumens) from one or more sources to a given area.

To cover the design and application of outdoor lighting adequately, a complete book, if not volumes, would be required; however, this section gives the reader an introduction to the basic principles of outdoor lighting. Illumination data for outdoor lighting fixtures are usually shown in the form of *isofootcandle graphs* as shown in Figure 7–12. These are regarded as the most useful form of data since they are actual representations of the lighting pattern and intensity at grace level on a horizontal plane. The curves on the graph are points of equal illumination (in lumens per square foot or footcandles) connected by a continuous line and are known as *isofootcandle* (IFC) or *isolux lines*.

Two types if isofootcandle graphs commonly used to express footcandle levels for outdoor lighting units. Type 1 (Figure 7–13) is based on a *fixed* mounting height where the grid lines of the graph indicate actual distance in feet from the lighting unit. Type 2 (Figure 7–14) is computed for a *given* mounting height; however, the grid lines are indicated as ratios of the actual distance to the mounting height. The graph in Figure 7–14, for example, is calculated on the basis of a mounting height of 16 feet. To obtain the distance at which any of the isolux lines occurs, simply multiply the 16 feet by the corresponding grid-line ratio

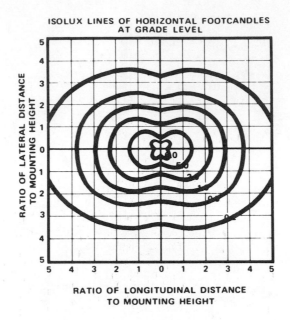

Figure 7–12   TYPICAL ISOFOOTCANDLE GRAPH.

number. For example, at what distance does the IFC occur along the longitudinal distance from the lighting unit? The IFC (isogrid line (1) is 16 feet since 16 feet multiplied by 1 equals 16 feet—the distance from the lighting unit.

The distance ratios allow *correction factors* (CF) to be used for other mounting heights. Correction factors for other mounting heights can be determined by dividing the square of the given mounting height by the square of the desired mounting height.

i.e.: given mounting height 16 feet
$$16 \times 16 = 256$$
desired mounting height 12 feet
$$12 \times 12 = 144$$

then CF = $\dfrac{265}{144}$ = 1.77

1.77 then becomes the correction factor to be applied to the footcandle level on the isofootcandle graph for a 12-foot mounting height.

The Panorama isofootcandle graphs are based on a single lighting unit. They become extremely useful in determining exact lighting levels at any specific point and for determining quickly required spacing in order to attain desired footcandle

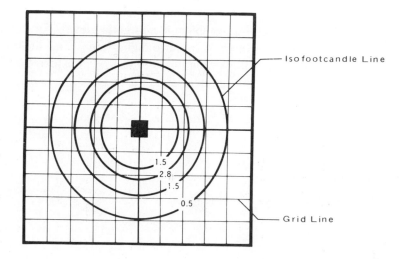

Figure 7–13   ISOFOOTCANDLE GRAPH BASED ON A FIXED MOUNTING HEIGHT.

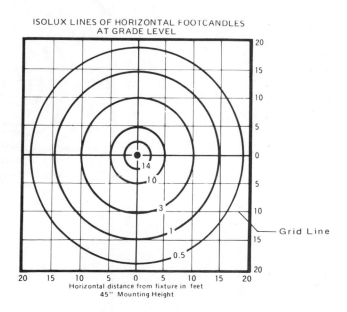

Figure 7–14   ISOFOOTCANDLE GRAPH COMPUTED FOR A GIVEN MOUNTING HEIGHT.

levels. These graphs can be quickly transferred to actual overlays of plot plans or layouts and light levels noted in Figure 7–15. Wherever isolux lines overlap, the two footcandle values are added together.

It is difficult to establish ideal lighting levels for outdoor spaces since much depends on the lighting in surrounding areas and the focal points available as reference. The following guidelines can be considered minimums for safety and emphasis in a dark ambient.

Outdoor Space	Footcandles
General lighting	0.5
Paths, walkways, steps	1
Backgrounds—fences, walks, trees, etc.	2
Flower beds, gardens	5
Trees, shrubbery—when emphasized	5
Focal points—large	10
Focal points—small	20
Parking areas	1–2
Pedestrial building entrances	2

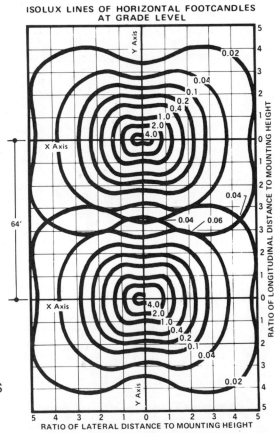

Figure 7–15 WHEREVER ISOLUX LINES OVERLAP, THE TWO FOOTCANDLE VALUES ARE ADDED TOGETHER.

LEVELS OF ILLUMINATION

	Recommended Footcandles (Minimum At Any Time)		Recommended Footcandles (Minimum At Any Time)
Building—		**Parking Lots**	5
General Construction	10	Self-Parking	1
Excavation Work	2	Attendant Parking	2
Building Exteriors and Monuments, Floodlighted—		**Piers, Freight and Passenger**	20
Bright Surroundings—		**Prison Yards**	5
Light Surfaces	15	**Quarries**	5
Dark Surfaces	50		
Dark Surroundings—		**Railroad Yards—Classification**	
Light Surfaces	5	Switch Points	2
Dark Surfaces	20	Body of Yard	1
Bulletins and Poster Boards— (Water Tanks or Stacks With Advertising Messages, Flags)		**Service Stations (At Grade)—**	
		Light Surroundings—	
Bright Surroundings—		Approach	3
Light Surfaces	50	Pump Island Area	30
Dark Surfaces	100	Service Areas	7
Dark Surroundings—		Dark Surroundings—	
Light Surfaces	20	Approach	1.5
Dark Surfaces	50	Pump Island Area	20
		Service Area	3
Coal Yards (Protective)	0.2	**Shipyards—**	
Dredging	2	General	5
		Ways	10
Loading Platforms	20	Fabrication Area	30
Lumber Yards	1	**Storage Yards, Active**	20

Figure 7–16 LEVELS OF ILLUMINATION FOR OUTDOOR LIGHTING APPLICATIONS. (COURTESY, WESTINGHOUSE)

The table in Figure 7–16 lists recommended illumination levels for many other outdoor applications. An outdoor lighting installation should be designed so that the illumination will not fall below these recommendations at any time during the maintenance cycle; therefore, an installation allowance must be made for reasonable depreciation. Also, the reflectance of the surface and the brightness of the surroundings must be considered in order to determine the amount of light necessary.

Methods of calculating outdoor lighting requirements are numerous, but the following steps are usually required regardless of the method used.

1   Determine the level of illumination required for the area in question. (See Figure 7–17.) Use the table in Figure 7–6 or similar tables from manufacturers' catalogs.

2   Select the type of lighting fixture and its type of light source i.e., fluorescent, incandescent, mercury vapor, H.I.D., etc. The study of lighting-fixture catalogs will help immensely in making the correct choice.

3   Determine the number of lumens in the beam of each light source (luminaire) in order to know how many fixtures (luminaires) will be required. Lighting-fixture data in manufacturers' catalogs normally give procedures for determining this with their particular type of fixtures.

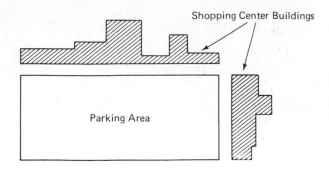

Figure 7–17   PLAN VIEW OF A SHOPPING CENTER PARKING AREA.

4   Estimate the maintenance factor (lamp-light level depreciation) due to dirt or life of lamp. Maintenance factors are usually estimated to be from 0.65 to 0.85 (65 to 85 percent).

5   Determine the number of lighting fixtures required by the equation:

$$\text{Number of Luminaires} = \frac{\text{area} \times \text{desired footcandles} \times WSF}{\text{Luminaire watts per lamp}} \qquad (7\text{–}4)$$

where:

$Area$ = surface to be lighted in square feet (ft²).

Desired footcandles = recommended illumination level of area to be illuminated,

Luminaire watts = total watts of lamp selected for given application,

WSF = a utilization factor which combines lamp lumens, beam efficiency, and maintenance factor.

6   Lay out the fixtures (luminaires) for coverage and uniformity. After a tentative layout has been made (Steps 1 to 5), the uniformity may be checked by calculating the intensity of illumination at a few points on the lighted surface. This may be done by using an isofootcandle graph as discussed previously.

Example 7–5

The floor plan in Figure 7–17 is that of a parking area measuring 240 feet × 480 feet for a small shopping center. It is desired to illuminate the area to meet current IES lighting standards for outdoor applications.

Solution

Step 1 Since this is a parking lot, the desired illumination level is 1 footcandle, as found in the table in Figure 7–16.

Step 2 Judgment is required to select the type of lighting fixture that has a good appearance by day, gives efficient performance at night, and blends tastefully with the architectural design of the shopping-center buildings. Let us choose 400-watt Mercury lamps as the lamps to try first. A General Electric, large-lamp catalog indicates that this type of lamp provides 20,500 initial lumens.

Step 3 Steps 3 and 4 given previously are combined in a utilization factor WSF (watts per square foot). A single WSF factor will not give accurate results for areas of all sizes; however, the four factors in the following table will suffice for most applications.

WSF FACTOR

Small area (1000 to 3000 square feet)	0.16
Medium area (3000 to 20,000 square feet)	0.11
Large area (20,000 to 80,000 square feet)	0.08
Extra-large area (over 80,000 square feet)	0.06

The above table shows that the WSF for the area in question ($240 \times 480 = 115{,}200$ square feet) will be 0.06. This means that 0.06 watt of "lamp power" is required to place 1 footcandle of illumination within 1 square foot of the area to illuminated.

Step 4 The number of lighting fixtures required for the area may be found by substituting known values in the WSF equation.

$$\frac{115{,}200 \times 1 \times 0.06}{400} = 17.28 \text{ lamps}$$

At this point, it is necessary to get an even number of lamps on each pole and an even number of poles so that the area will be illuminated uniformly. It was decided that 10 poles with two 400-watt lamps on each would give the desired illumination with a small amount to spare.

Step 5 The fixtures may be laid out as shown in Figure 7–18. This is only one of several possibilities, however.

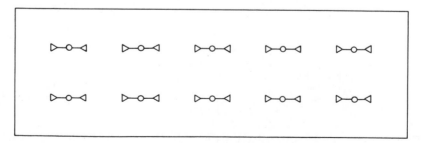

Figure 7–18 PARKING LOT AREA WITH FIXTURES LAID OUT.

7–6   LIGHTING CONTROLS

Many control devices have been developed to make the best use of lighting equipment. Automatic timing devices for turning lights on and off at various intervals are used extensively for outdoor lighting. Dimmers are used to vary the intensity of light sources. The common wall switch is used in almost every lighting application to turn lights on and off manually.

Wall switches used on branch circuits to control lighting will usually fall into one of the three basic categories:

1   Snap-action switches

2   Quiet switches

3   Mercury switches

A snap-action switch consists of a device containing two stationary, current-carrying elements, a moving current-carrying element, a handle for the moving element, a spring, and an enclosure. When the switch is properly installed and the handle is in the down or OFF position, no current can pass, and the light or lights controlled by the switch will not operate. When the moving element is closed by switching the handle to the up or ON position, the circuit is complete, and the light will ''burn.'' This action is pictured in Figure 7–19.

Mercury switches consist of a sealed capsule containing mercury. A handle is provided to tilt the capsule to allow the mercury to float to one end to bridge two contact points when the switch is in the ON position, and to tilt the mercury away from the contact points when the switch is in the off position. Such switches offer the ultimate in silent operation but are much higher in price than either the snap-action or quiet switch.

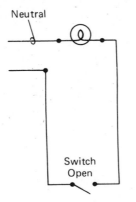

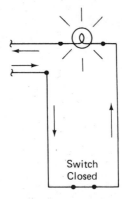

Figure 7–19 DIAGRAM SHOWING   Figure 7–19b WIRING DIAGRAM
WALL SWITCH IN OFF POSITION.   SHOWING SWITCH IN ON POSITION.

The quiet switch is a compromise between the snap-action switch and the mercury switch. Its operation is considerably quieter than the snap-action switch, but it is not as expensive as the mercury switch. It is the most commonly used switch in modern lighting systems and is manufactured for loads from 10 to 220 amps, in single-pole, two-pole, three-way, four-way, etc.

Three-way switches are used to control one or more lamps from two different locations, such as at the top and bottom of stairways, in halls, and similar places. Unlike single-pole switches that control a light or group of lights from only one location (containing two wire terminals), the three-way switches make it possible to control lighting from two locations, as can be seen in the wiring diagram in Figure 7–20. By tracing the circuit, we may see how these three-way switches operate. Two wires are connected to the 120-volt line; one wire is connected directly to the light fixture (this is the neutral or white wire); the other (black or "hot") wire continues on to one of the three-way switches. If both handles of the three-way switches are in the "up" position, the current will pass through the top "traveler" wire (between the switches) and on through the other switch to the lamp, which will light because of the completed circuit. If either of the handles is turned to the "down" position, the circuit will be opened, and the lamp will go out. However, the lamp may be turned on again by changing the position of either handle of either switch; that is, if the handle of the three-way switch is turned to the "down" position, the light will burn, or it will burn if the handle on the right-hand, three-way switch is turned to the "up" position. Thus, the lamp is controlled by switches at two locations.

In order to control a light or a group of lights from more than two locations, one or more four-way switches will have to be added to the three-way switches. In circuits of this type, two three-way switches will always be required—one on each side of the group of four-way switches—and one four-way switch for each additional location. For an illustration, look at the wiring diagram in Figure 7–21, which shows a lamp controlled from four locations.

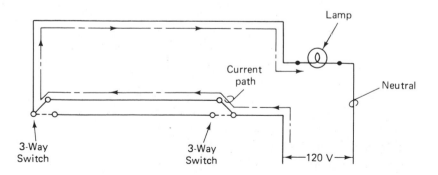

Figure 7–20　WIRING DIAGRAM OF A THREE-WAY SWITCH CIRCUIT.

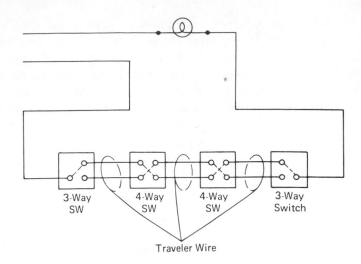

3-Way   4-Way   4-Way   3-Way
SW   SW   SW   Switch

Traveler Wire

Figure 7–21   WIRING DIAGRAM SHOWING A LAMP CONTROLLED FROM FOUR DIFFERENT LOCATIONS.

In general, the wiring diagram shows a 120-volt, two-wire circuit feeding a lighting outlet. The white or neutral wires connect directly to the terminal on the lighting fixture while the black wire continues to one of the three terminals on one of the three-way switches. This terminal is known as the "point" terminal. Two "traveler" wires are then connected to the remaining terminals and are run to two of the four terminals on the next four-way switch. Two other "traveler" wires are connected to the other two terminals on this four-way switch and are run to the other four-way switch. The two remaining "traveler" wires connect to the other three-way switch, and one wire from the "point" terminal connects to the other side of the lighting fixtures. The actuation of any one of these four switches will turn the light on or off.

7–7   UTILIZING TIME SWITCHES FOR LIGHTING CONTROL

Time controls are used extensively for the control of parking-lot lighting, security lighting, blinker light control (in school zones), and similar applications. In installations, such as shopping centers, small factories, or garden apartment projects, where future expansion is anticipated, the most versatile and inexpensive time control is a 4-pole unit. Available 4-pole, time-control units include a 24-hour general purpose unit, a 24-hour Astro dial, and a 7-day general purpose unit. Each of these controls can handle four different lighting circuits or can control three phase loads. One or two circuits can be used in the original installation, and the additional capacity is available as needed. When new circuits are added, they will be synchronized with the existing lights.

A wiring diagram for a typical 4-pole, single-throw time switch is shown in Figure 7–22. Note that circuits 1, 2, 3, and 4 connect to terminals 8, 6, 4, and 2, respectively. The loads are connected to 1, 3, 5, and 7 and are controlled simultaneously by the time control at preselected intervals.

Another control device that can be used to control lighting, either alone or when used in conjunction with a time switch, is the photocontrol. Such a control turns lights ON at dusk and OFF at dawn. The connection is made similar to a standard, single-pole wall switch, as shown in Figure 7–23.

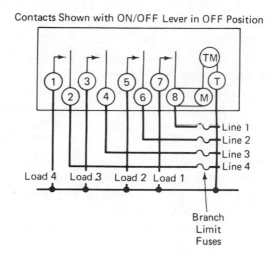

Figure 7–22   WIRING DIAGRAM FOR A TYPICAL FOUR-POLE, SINGLE-TIME SWITCH.

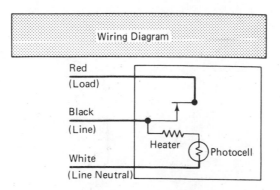

Figure 7–23   WIRING DIAGRAM OF A PHOTOELECTRIC CONTROL CONNECTION.

7–8   DIMMING LIGHTS CIRCUITS

Many modern lighting applications make it desirable to provide a means of varying the intensity of the illumination for a given area. For example, the dimming lights in the residential dining area creates a mood of candlelight for formal dining; dimming recreation room lights provides a better atmosphere for viewing TV or for engaging in casual conversation; a hall light dimmer at night serves as a night light without using excessive current.

Dimming incandescent lamps is best accomplished by reducing the voltage by means of a variable transformer, magnetic amplifier, saturable reactor, or electronic circuit. Most dimmers are manufactured in various ratings from 300 to 2000 watts.

For dimming conventional lighting circuits, most dimmers can be used to replace a regular wall switch by merely disconnecting the two leads from the switch and reconnecting the two wires to the screw terminals on the dimmer as shown in Figure 7–24. The dimmer should be of sufficient capacity to handle the lighting load installed on the circuit. For all practical purposes, the installation of modern lighting dimmers is performed in much the same way as the installation of a conventional wall switch.

Dimmers may also be installed on three-way switch systems, but dimming can be accomplished only at the point where the dimmer is located. However, the lamp load may be turned on and off at any point in the dimming cycle at the dimmer itself or at the location of the other three-way switch. Under no circumstances should two three-way dimmers be wired together on the same circuit; this will result in improper dimming. The dimmer must also be used in conjunction with the conventional three-way or four-way switch—a wiring of which is shown in Figure 7–25.

Typical Installation in Single Gang Box

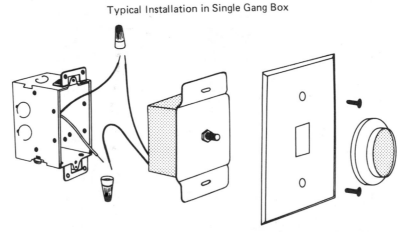

Figure 7–24   METHOD OF CONNECTING A DIMMER CONTROL IN A WALL SWITCH BOX.

The advent of the rapid-start principle of fluorescent lamp operation made possible the dimming of fluorescent lamps. A cathode-heating supply current is provided with a constant voltage while, at the same time, the arc passing through the tube is varied to produce dimming. In order to maintain this constant cathode-heating current while varying the lamp current, a special ballast transformer is necessary. It should be connected to the dimmer circuit, as shown in Figure 7–26.

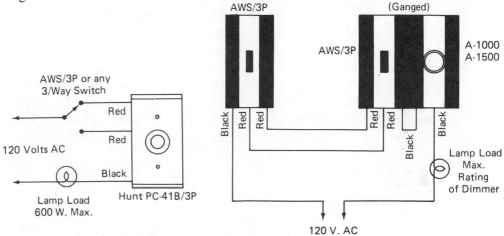

Figure 7–25   WIRING DIAGRAM SHOWING A DIMMER USED IN CONJUNCTION WITH TWO THREE-WAY SWITCHES.

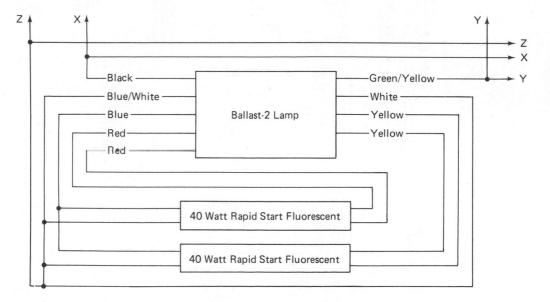

Figure 7–26   WIRING DIAGRAM OF A DIMMING SYSTEM FOR FLUORESCENT LAMPS.

A 120-volt line voltage is fed into the primary of the ballast transformer in Figure 7–26; this produces constant voltage on the secondary winding that is connected to the cathode of the fluorescent lamp. The current that passes the length of the lamp, however, must pass through the main secondary of the ballast transformer and also through the dimmer. Leads connect to other ballasts (controlling other lamps) in the system. Dimming ballasts of this type are commonly available for 40-watt, rapid-start, fluorescent lamps.

Complete, semiautomated, dimming systems are usually preferred for cinemas and other auditoriums. Such a system enables the projectionist to touch a switch momentarily and to have the lights automatically raise or lower. A typical application would be for a four-circuit auditorium and three-circuit, screen-lighting system. However, circuits may be added to this arrangement, up to a maximum of nine per single housing. Circuits can also be deleted if they are not required. A wiring diagram for this application appears in Figure 7–27.

In general, the dimming arrangement consists of four circuits to control

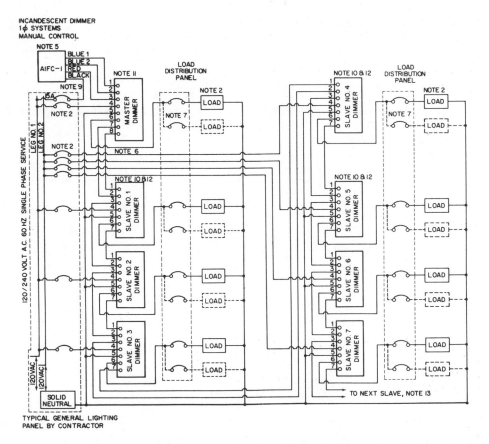

Figure 7–27   WIRING DIAGRAM OF A LIGHTING DIMMING SYSTEM FOR AN AUDITORIUM.

auditorium lights. Each circuit may carry a load of 2000 watts of incandescent lighting at 120 volts; each circuit is then fused at 20 amperes. A motorized drive, which takes approximately 16 seconds to raise or lower the lights when special circuits are momentarily energized by the projectionist, is incorporated into the system. Emergency switches should also be provided at other remote locations. When activated, the auditorium lights will immediately go to full bright.

Three circuits are also provided for screen lights. Again, each circuit can carry a maximum of 2000-watts incandescent lighting, and each should be fused at 20 amperes. Special circuits are incorporated into this system which takes control of the motorized drive—when they are momentarily actuated by the projectionist—and raises and lowers the screen lights. The full cycle takes about 16 seconds although almost any time desired is available. The input power for the circuit in Figure 7–27 is designed for either 120/240 single phase or two phases of 120/208 volts.

QUESTIONS

Answer the following questions by filling in the blanks.

1   Although there are several methods in which lighting calculations may be made, most electricians use the _____ method to calculate illumination requirements for areas in the home.

2   In laying out a lighting installation, select lighting fixtures that give approximately _____ lumens in the kitchen area.

3   Most people consider a _____ ceiling in the kitchen ideal for general lighting.

4   The _____ method of lighting calculations is normally used to determine the average maintained illumination levels on a given work plane.

5   The maintenance factor of a dusty concrete plant probably is 0._____ .

6   Wall switches used to control lighting circuits usually fall into one of the following basic categories: _____, _____, and _____.

7   In the diagram in Figure 7–28, connect the terminals on the three-way switches with a line (to indicate conductors) to make the three-way switches operate properly.

8   The dimmer for a particular lighting circuit should be of _____ capacity to handle the lighting load.

9   Most dimmer control switches are rated from _____ watts to _____ watts.

10   Basically, outdoor lighting is simply a matter of delivering a sufficient quantity of _____ from one or more light sources to a given area.

11   If the squares on the isolux curve in Figure 7–29 are 10 feet × 10 feet, the footcandle level directly in front of the pole, 30 feet away is, _____.

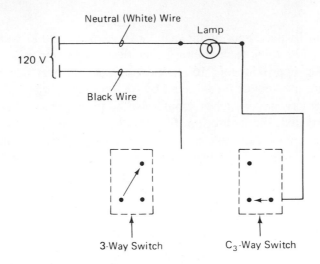

Figure 7–28 WIRING DIAGRAM FOR ASSIGNMENT.

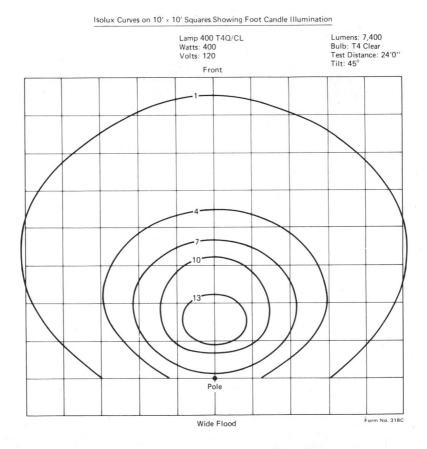

Figure 7–29 ISOLUX CURVES FOR USE IN ASSIGNMENT.

PROBLEMS

A residential kitchen measures 17.7 feet in one direction and 11.3 feet in another. Calculate:

a   The total area in square feet.

b   The total recommended lumens of illumination (use Table 7–1).

In the cross-sectional view of the room in Figure 7–30, calculate:

a   The ceiling-cavity ratio.

b   The floor-cavity ratio.

c   The room-cavity ratio.

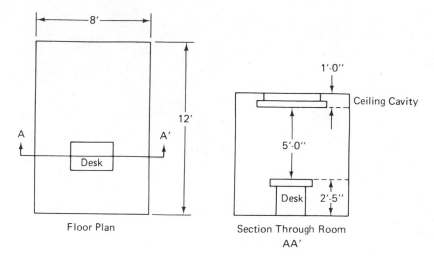

Floor Plan   Section Through Room
                           AA'

Figure 7–30

8

BRANCH CIRCUIT INSTALLATION

Most of the material in this chapter deals with the portion of the electrical system involving items such as outlet boxes of all types, all 1 inch and smaller conduits with fittings, and types of cable and building wire up to and including No. 8 AWG. In general, branch circuits include all wiring for lighting, receptacles, and small-power and communication systems.

Specific items covered include branch circuit rough-in, surface metal raceway systems, branch circuit wiring, busway, and branch circle cable.

8–1 OUTLET BOXES AND FITTINGS

Outlet boxes, cabinets, and raceway fittings are normally made of steel, although aluminum or brass is used to some extent. They are also made of porcelain or plastic materials for use in nonmetallic wiring systems, but they have not been widely used because of the higher cost and breakage.

When conduit systems are installed in a building, the entire plan should be carefully analyzed to make certain that the proper size conduit has been selected to contain the required number and size of conductors; that junction boxes are of the proper size and are located in a readily accessible place; and that conduit runs, including homeruns to panelboards, have been routed in the shortest way possible in order to save conduit and wire, as well as labor. The location of outlet boxes in concrete structures is especially important since an error or omission can be extremely costly to correct after the concrete is poured.

The installation of the complete conduit system is termed "roughing in." Only the conduit, outlet boxes, and other mechanical devices are installed; none of the wires is pulled in until the conduit system is complete. On larger projects, it may take weeks or months to complete the conduit system before the wires may be pulled in; therefore, it is very important to have complete and accurate working drawings for any project. Any deviation from the original drawings should be recorded by sketches, written notes, or both.

8–2   OUTLET BOXES

Electrical drawings rarely indicate the exact types and sizes of outlet boxes to be used in a given location, with the possible exception of hazardous areas. Therefore, electricians must have a good knowledge of all types of outlet boxes and be able to select the best types (and the correct sizes) for any given application. For example, outlet boxes for use with type AC (armored cable) cable should contain built-in cable clamps specifically designed for type AC cable.

The time involved during the installation of raceway or cable systems can be greatly reduced if the proper outlet boxes are used. The chart in Figure 8–1 gives several types of outlet boxes currently available. Electricians unfamiliar with the numerous types of other outlet boxes should obtain current catalogs from the outlet box manufacturers and study the illustrations and descriptions. With this basic information, on-the-job experience will go smoother.

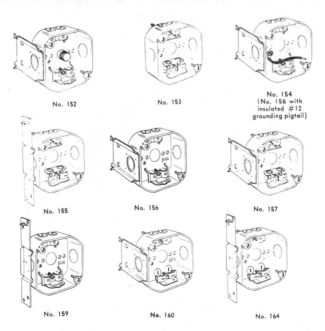

No. 152   No. 153   No. 154
(No. 156 with
insulated #12
grounding pigtail)

No. 155   No. 156   No. 157

No. 159   No. 160   No. 164

Figure 8–1   CHART GIVING SEVERAL TYPES OF OUTLET BOXES CURRENTLY AVAILABLE.

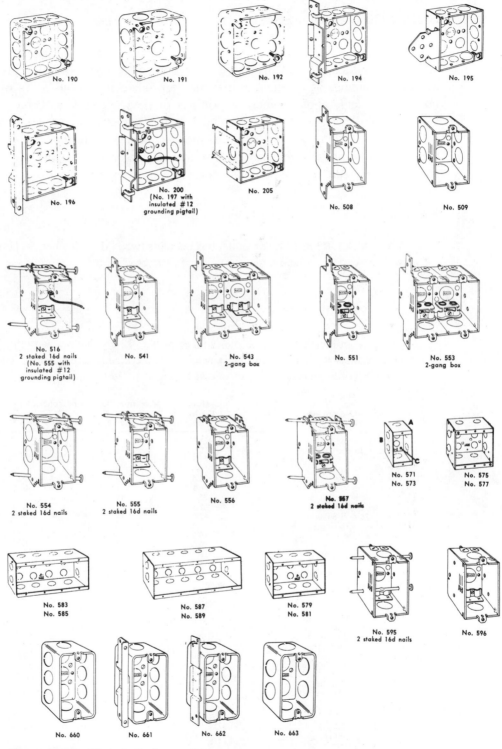

No. 190

No. 191

No. 192

No. 194

No. 195

No. 196

No. 200
(No. 197 with
insulated #12
grounding pigtail)

No. 205

No. 508

No. 509

No. 516
2 staked 16d nails
(No. 555 with
insulated #12
grounding pigtail)

No. 541

No. 543
2-gang box

No. 551

No. 553
2-gang box

No. 554
2 staked 16d nails

No. 555
2 staked 16d nails

No. 556

No. 557
2 staked 16d nails

No. 571
No. 573

No. 575
No. 577

No. 583
No. 585

No. 587
No. 589

No. 579
No. 581

No. 595
2 staked 16d nails

No. 596

No. 660

No. 661

No. 662

No. 663

Figure 8–1   (Concluded)

8–3 SIZING OUTLET BOXES

The NEC lists specific instructions for sizing outlet boxes for the number and sizes of conductors entering the box. Figure 8–2, for example, gives the maximum number of conductors that may terminate in various sizes of outlet boxes. This table applies where no fittings or devices, such as cable clamps, switches, and receptacles, are contained in the box and where no grounding conductors are part of the wiring within the box. When one or more cable clamps or fixture studs are contained in the box, the number of conductors given in the table must be reduced by one conductor.

Example 8–1

A 4 inch × 1-1/2 inch square outlet box contains two NM cable clamps. How many No. 12 AWG conductors may terminate in the box?

Solution

The table in Figure 8–2 gives the number of No. 12 conductors as eight, but this number must be reduced by one since NM cable clamps are present. Therefore, seven No. 12 conductors are permitted to terminate in the box.

Box Dimension, Inches Trade Size or Type	Min. Cu. In. Cap.	Maximum Number of Conductors				
		No.14	No.12	No.10	No.8	No.6
4 x 1¼ Round or Octagonal	12.5	6	5	5	4	0
4 x 1½ Round or Octagonal	15.5	7	6	6	5	0
4 x 2⅛ Round or Octagonal	21.5	10	9	8	7	0
4 x 1¼ Square	18.0	9	8	7	6	0
4 x 1½ Square	21.0	10	9	8	7	0
4 x 2⅛ Square	30.3	15	13	12	10	6*
4 11/16 x 1¼ Square	25.5	12	11	10	8	0
4 11/16 x 1½ Square	29.5	14	13	11	9	0
4 11/16 x 2⅛ Square	42.0	21	18	16	14	6
3 x 2 x 1½ Device	7.5	3	3	3	2	0
3 x 2 x 2 Device	10.0	5	4	4	3	0
3 x 2 x 2¼ Device	10.5	5	4	4	3	0
3 x 2 x 2½ Device	12.5	6	5	5	4	0
3 x 2 x 2¾ Device	14.0	7	6	5	4	0
3 x 2 x 3½ Device	18.0	9	8	7	6	0
4 x 2⅛ x 1½ Device	10.3	5	4	4	3	0
4 x 2⅛ x 1⅞ Device	13.0	6	5	5	4	0
4 x 2⅛ x 2⅛ Device	14.5	7	6	5	4	0
3¾ x 2 x 2½ Masonry Box/Gang	14.0	7	6	5	4	0
3¾ x 2 x 3½ Masonry Box/Gang	21.0	10	9	8	7	0
FS—Minimum Internal Depth 1¾ Single Cover/Gang	13.5	6	6	5	4	0
FD—Minimum Internal Depth 2⅜ Single Cover/Gang	18.0	9	8	7	6	3
FS—Minimum Internal Depth 1¾ Multiple Cover/Gang	18.0	9	8	7	6	0
FD—Minimum Internal Depth 2⅜ Multiple Cover/Gang	24.0	12	10	9	8	4

Figure 8–2 THE MAXIMUM NUMBER OF CONDUCTORS THAT MAY TERMINATE IN VARIOUS SIZES OF OUTLET BOXES.

Volume Required Per Conductor

Size of Conductor	Free Space Within Box for Each Conductor
No. 14	2.0 cubic inches
No. 12	2.25 cubic inches
No. 10	2.5 cubic inches
No. 8	3.0 cubic inches
No. 6	5.0 cubic inches

Figure 8–3 THE VOLUME PER CONDUCTOR ALLOWED FOR VARIOUS OUTLET BOXES WHEN A COMBINATION OF CONDUCTOR SIZES ARE USED.

One conductor shall also be deducted from the number in the table for each switch, receptacle, or other wiring device attached to the box. A further deduction of one conductor must be made for one or more grounding conductors.

Example 8–2

A 3 inch × 2 inch × 2-3/4 inch deep box contains two NM cable clamps where two No. 12-2 w/gd. cable terminates to feed a duplex receptacle. Is the box of sufficient size to accommodate the components?

Solution

The table in Figure 8–2 gives six as the number of No. 12 conductors permitted in the size box with no clamps, wiring devices, etc. One conductor must be deducted for the cable clamp, one for the grounding wires, and one for the wiring device. This leaves a total of three conductors permitted in the box. Since the two 12-2 w/gd. cables contain a total of four conductors, the box is too small for the application. A 3 inch × 2 inch × 3-1/2 inch switch box will have to be used.

For combinations of conductor sizes shown in Figure 8–2, the volume per conductor in the table in Figure 8–3 shall apply. However, the maximum number and size of conductors listed in the table in Figure 8–2 must not be exceeded.

8–4 INSTALLATION OF OUTLET BOXES

Outlet boxes must be securely fastened in place, unless otherwise provided for specific purposes in the NEC. Boxes attached to wood studs are normally secured with nails; boxes attached to metal studs are secured with screw-metal screws; boxes attached to metal beams are normally secured with nuts and bolts or welded in place; boxes attached to masonry walls are secured with masonry anchors (Figure 8–4), and so forth.

Article 370–13 of the NEC requires the following procedures for mounting outlet boxes: Boxes shall be securely and rigidly fastened to the surface upon which they are mounted or securely and rigidly embedded in concrete or masonry. Where nails are used as a mounting means and pass through the interior of the box, they shall not be more than 1/4 inch from the back of the box. Boxes shall be supported from a structural member of the building either directly, by using a substantial and approved metal or wooden brace, or as otherwise provided in this section. If made of wood, the brace shall not be less than nominal one-inch thickness. If made of metal, it shall be corrosion resistant and not be less than No. 24 MSG.

Where mounted in new walls in which no structural members are provided or in existing walls in previously occupied buildings, boxes not over 100 cubic inches in size, specifically approved for the purpose shall be affixed with approved anchors or clamps so as to provide a rigid and secure installation. Threaded boxes or fittings not over 100 cubic inches in size that do not contain devices or support fixtures shall be considered adequately supported if two or more conduits are threaded into the box wrench tight and are supported within three feet of the box on two or more sides, as is required by this section. Threaded boxes or fittings not over 100 cubic inches in size shall be considered to be adequately supported if two or more conduits are threaded into the box wrench tight and are supported within 18 inches of the box, as required by this section.

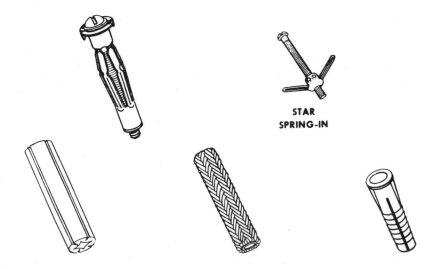

STAR
SPRING-IN

Figure 8–4   MASONRY ANCHORS USED TO SECURE OUTLET BOXES TO MASONRY WALLS.

8–5   CABLE WIRING

Nonmetallic-sheathed (type NM) and armored (type AC) cables are very popular for use in residential and small commercial wiring systems. In general, both types of cable may be used for both exposed and concealed work in normally dry locations. They shall be installed or fished in air voids in masonry block or tile walls where such walls are not exposed or subject to excessive moisture or dampness.

Type NM cable shall not be installed where exposed to corrosive fumes or vapors; nor imbedded in masonry, concrete, fill or plaster; nor run in shallow chase in masonry or concrete and covered with plaster of similar finish. This cable must not be used as a service-entrance cable, in commercial garages, theaters and assembly halls, motion-picture studios, storage battery rooms, hoistways, hazardous locations, or embedded in poured cement, concrete, or aggregate. Type AC cable has the same restrictions.

For use in wood structures, holes are bored through wood studs and joists, and the cable is then pulled through these holes to the various outlets. The holes normally give sufficient support, providing they are not over four feet on center. Where no stud or joist support is available staples or some similar supports are required for the cable. The supports must not exceed 4-1/2 feet and must be within 12 inches of each outlet box or other termination point.

Proper tools facilitate the running of branch circuit cables and include items such as sheathing strippers for stripping the NM cable (Figure 8–5), hacksaw for cutting and stripping type AC (BX) cable (Figure 8–6), carpenter's apron for holding staples, crimp connectors, and wire nuts.

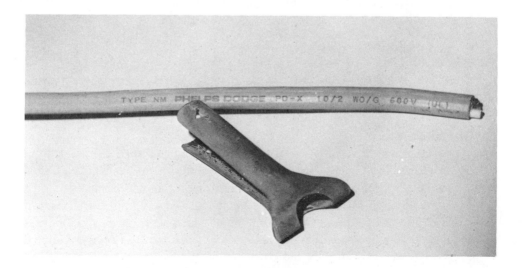

Figure 8–5   SHEATHING STRIPPERS FOR STRIPPING TYPE NM CABLE.

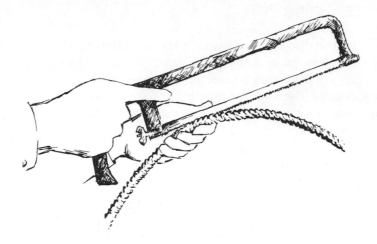

Figure 8–6   HACKSAW FOR CUTTING AND STRIPPING TYPE AC CABLE.

8–6   RACEWAY INSTALLATIONS

In general, an electrical raceway system is designed expressly for holding conductors, and includes, in addition to rigid, EMT, PVC, and other types of conduit, the outlet boxes and other fittings through which the conductors of the system will be installed. For general building construction, rigid or PVC conduit is normally used in and under concrete slabs; EMT is used for all above-surface installations except where the system will be exposed to severe mechanical injury.

Rigid Metal Conduit

Galvanized rigid metal conduit may be used under all atmospheric conditions and occupancies. However, unless corrosion protection is provided, rigid metal conduit and fittings should not be installed in areas subject to severe corrosive influences.

During the installation of rigid conduit, it has been the practice to use special conduit cutters to cut the conduit. These cutters normally leave a large burr, and often a definite hump, inside the conduit, requiring additional time to remove the burr. A better method is to use a lightweight, portable electric saw using blades with 18 teeth per inch. Conduit cuts should be made square, and the inside edge of the cut should be adequately reamed to remove any burr or sharp edge that might damage the insulation of the conductors when pulled in later. Lengths of conduit should be accurately measured before they are cut since recutting obviously results in lost time.

The contractor normally provides for each crew a vice stand that will securely hold the conduit and not shift about when the cut is being made. A power hacksaw of either the blade or band type should be provided, as well as a sufficient supply of hacksaw blades to be used in the electrician's hacksaw frames. The cut should be made entirely through the conduit, not broken off the last fraction of an inch. While the hacksaw may be used to cut smaller sizes of conduit by hand, the larger sizes should not be cut by hand except in emergencies. Hand cutting of the large sizes takes up too much time, and it is extremely difficult to cut such large sizes of conduit square.

In addition to the length needed for the piece of conduit for a given run, allow an additional 3/8 inch on smaller sizes of conduit for the wall of the box and the bushing. Because larger sizes of conduit are usually connected to heavier boxes, allow approximately 1/2 inch. If additional locknuts are needed in a run, a 1/8-inch allowance for each locknut will be sufficient. Where conduit bodies are used, the length of the threaded hub should be included in the measurements.

The usual practice for threading the smaller sizes of rigid conduit is to use a pipe vice in conjunction with a die stock with proper size guides and sharp cutting dies properly adjusted and securely held in the stock. Clean, sharp threads can be cut only when the conduit is well lubricated; use a good lubricant and plenty of it.

Conduits should always be cut with a full thread. To accomplish this, run the die up on the conduit until the conduit comes through the die for about one full thread. This gives a thread length that is adequate for most purposes. However, do not overdo it; if the thread is too long, the portion that does not fit into the coupling will erode because threading removes the protective covering. Clean, sharply cut threads make a better continuous ground and save much grief and unnecessary labor. It takes a little extra time to make certain that threads are properly made, but a little extra time spent at the beginning of a job may save much time later on.

On projects where a large number of relatively short sections of conduit are required for nipples, considerable threading time can be saved by periodically gathering up the short lengths of conduit resulting from previous cuts and reaming and running a thread on one end with a power threader and redistributing these lengths to the installation points about the job. This procedure eliminates one-hand threading operation in many instances.

When threadless couplings and connectors are used with rigid conduit, they should be made tight. Where the conduit is to be buried in concrete, the couplings and connectors must be of the concrete type; where the conduit is used in wet locations, they must be of the rain-tight type.

The installation of rigid conduit for branch circuit raceways often requires many changes of direction in the runs ranging from simple offsets to complicated angular offsets, saddles, and so on. In bending elbows, care should be taken to comply with the NEC. In general, the Code states that an elbow or 90° bend must have a minimum radius of six times the inside diameter of the conduit. Therefore, the radius of two-inch conduit must have a radius of at least 12 inches, a three-inch conduit 18 inches, and so forth.

Bends are normally made in the smaller sizes of conduit by hand with the use of conduit hickeys or benders. In some cases, where many bends of the same type must be made, hand roller "Chicago benders" or hydraulic benders are used to simplify making the bends to certain dimensions.

Occasionally, rigid conduit will have to be rebent after it is installed. In such cases, the rebending must be done carefully so that the conduit does not break. Most often, these rebends will have to be made at stub ups—conduits emerging through concrete floors. To rebend conduit, the concrete should be chipped away a few inches around the conduit, and the conduit should be warmed with a propane torch. It can be bent into the required shape without further trouble.

Electrical Metallic Tubing

Electrical metallic tubing (EMT) may be used for both exposed and concealed work, except where the tubing will be subjected to severe physical damage and in cinder concrete, unless the tubing is at least 18 inches under the fill. The tubing should be cut with a hand or power hacksaw, using blades with 32 teeth per inch. The cut ends should be reamed to remove all rough edges. Threadless couplings and connectors must be made tight, and the proper type should be used for the situation; that is, concrete-tight types should be used when the tubing is buried in concrete, rain-tight type used when installed in a wet location, etc. Supports must be provided when installed above grade at least every 10 feet and within 3 feet of each outlet box or other termination point.

Bends are made in EMT much the same as in rigid conduit. However, roll-type benders are used exclusively for EMT. This type of bender has high supporting sidewalls to prevent flattening or kinking of the tubing and a long arc that permits the making of 90° bends (or any lesser bends) in a single sweep.

Many time-saving tools and devices have appeared on the market during the past years to facilitate the installation of EMT. Table hydraulic speed benders, for example, make 90° bends or offsets in 5 to 10 seconds. Shoes are available for 1/2-inch through 1-1/4-inch EMT. Mechanical benders are also available for sizes through two-inch EMT.

An automatic "kinker" will eliminate the need for offset connectors wherever 3/4-inch or 1/2-inch EMT is used. The end of a piece of EMT is inserted in the blocks of the device, and one push of the handle (about two seconds duration) makes a perfect offset. Every bend is identical, which eliminates time lost refitting or cutting and trying.

EMT hand benders with built-in degree indicators allow the operator to make accurate bends from 15° to 90° faster since in-between measurements are eliminated.

Surface Metal Raceway

Surface metal raceway is one of the exposed wiring systems that is quite extensively used in existing buildings while new wiring or extensions to the old are installed. Although it does not afford as rugged and safe a protection to the conductors as does rigid conduit or EMT, it is a very economical and quite dependable system when used under the conditions for which it was designed. The main advantage of surface metal raceway is its neat appearance where wiring must be run on room surfaces in finished areas.

Surface metal raceways may be installed in dry locations, except where subject to severe physical damage or where the voltage is 300 volts or more between conductors (unless the thickness of the metal is not less than 0.040 inch). Surface metal raceways should not be used in areas that are subjected to corrosive vapors, in hoistways, or in any hazard location. In most cases, this system should not be used for concealed wiring.

Various types of fittings for couplings, corner turns, elbows, outlets, and so on are provided to fit these raceways. Figure 8–7 shows a number of the most common fittings.

Many of the rules for other wiring systems also apply to surface metal raceways. The system must be continuous from outlet to outlet; it must be grounded; all wires of the circuit must be contained in one raceway, and so on.

In planning the installation of a surface metal raceway system, the electrical contractor should make certain that all materials are provided before work is begun. One missing fitting could hold up the entire project.

Proper tools should be provided to make the installation easier. For example, a surface metal raceway bending tool enables electricians to bend certain sizes of the raceway like rigid conduit or EMT. With such a tool, many of the time-consuming fittings can be eliminated since the tool can make 90° bends, saddles, offsets, back-to-back bends, and so on.

8–7 BRANCH CIRCUIT WIRING

In most cases, the installation of branch circuit wires is merely routine. However, there are certain practices that can reduce labor and materials to such an extent that such practices should at least be given careful consideration. The proper size and length of fish tape, as well as the type, should be one of the first considerations. For example, if most of the runs between outlets are only 20 feet or less, a short fish tape of, say, 25 feet will easily handle the job and will not have the weight and bulk of a larger one. When longer runs are encountered, the required length of fish tape should be enclosed in one of the metal or plastic fish tape reels so that the tape can be rewound on the reel as the play is being made in order to avoid having an excessive length of tape lying around on the floor. When

CATALOG NUMBER AND DESCRIPTION
211 90° FLAT ELBOW Base, each leg 1¼″ L. without tongue
211LH INTERNAL TWISTED ELBOW Base, each leg 2″ L. without tongue
211RH INTERNAL TWISTED ELBOW Base, each leg 2″ L. without tongue
214 PULL BOX 5″ L.
217 INTERNAL ELBOW Base, each leg 3″ L. without tongue
218 EXTERNAL ELBOW Base, each leg ⅞″ L. without tongue
228 ADJUSTABLE JUNCTION BOX 2½″ L., ¾″ W., ⅞″ D.
251 EXTENSION ADAPTER 4¾″ L., 3″ W., ½″ D.
289 REDUCING CONNECTOR 1⅞″ L. not including tongue
289A ADAPTER Connects Wiremold 200 to all 5700 series fittings
299 EXTENSION CONNECTOR

502 BUSHING
504 STRAP (1- or 2-hole)
506 CONNECTION COVER
511 90° FLAT ELBOW Base, each leg 2″ L. without tongue

CATALOG NUMBER AND DESCRIPTION
512 45° FLAT ELBOW Base, each leg ⅞″ L. without tongue
5514 LAY-IN FITTING 18″ L. Interconnects most 500 and 5700 fittings
517 INTERNAL ELBOW Base, each leg 2¾″ L. without tongue
5517C INTERNAL CORNER COUPLING For use with 5514
518 EXTERNAL ELBOW Base, each leg 2⅛″ L. without tongue
599 CONNECTOR

702 BUSHING
704 STRAP (1- or 2-hole)
706 CONNECTION COVER
711 90° FLAT ELBOW Base, each leg 2″ L. without tongue
712 45° FLAT ELBOW Base, each leg 1″ L. without tongue
717 INTERNAL ELBOW Base, each leg 2¾″ L. without tongue
718 EXTERNAL ELBOW Base, each leg 2⅛″ L. without tongue

57240 SINGLE POLE SWITCH AND BOX 4⅛″ L., 2″ W., 1⅜″ D., 10A, 125V; 5A, 250V
57242 UTILITY BOX 4⅛″ L., 2″ W., 1⅜″ D.
57243G DUPLEX GROUNDING RECEPTACLE AND BOX A.S.A. Standard, 4⅛″ L., 2″ W., 1⅜″ D. 15A, 125V

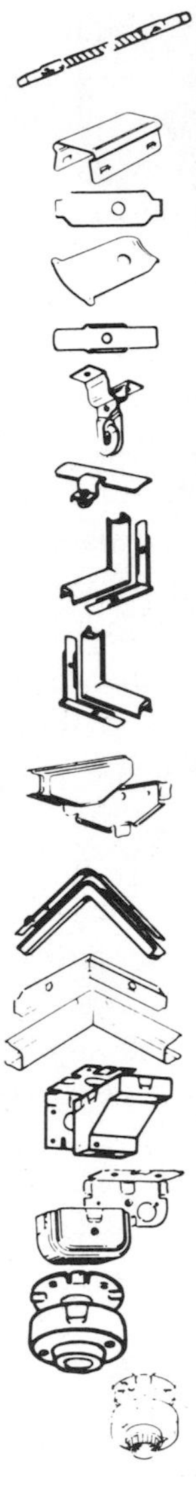

CATALOG NUMBER AND DESCRIPTION
5700F FLEXIBLE SECTION 18″ L. overall
5700WC* WIRE CLIP 1″ L.
5701 COUPLING
5701A* ADAPTER
5703 SUPPORTING CLIP 2½″ L.
5708 FIXTURE HOOK
5709 GROUND CLAMP
5711LH INTERNAL TWISTED ELBOW Base, each leg 2½″ L. without tongue
5711RH INTERNAL TWISTED ELBOW Base, each leg 2½″ L. without tongue
5715 TEE 3¾″ L., 1¾″ W., ⅞″ D.
5717A INTERNAL PULL ELBOW Base, each leg 5½″ L. without tongue
5717C* INTERNAL ELBOW
5719 CORNER BOX 2½″ L., 2⅜″ W., 2½″ D.
5719D CORNER BOX 4¹⁄₁₆″ L., 2½″ W., 1⅛″ D.
5721 UTILITY BOX 3″ Dia., 1¼″ D.
5726P KEYLESS RECEPTACLE 660W, 250V 3″ Dia., 2″ D.

*For use with 5700 Raceway.

Figure 8–7 SURFACE METAL RACEWAY FITTINGS.

several bends are present in the raceway system, the insertion of the fish may be made easier by using flexible fish tape leaders on the end of the fish tape.

The combination blower and vacuum fish tape systems are ideal for use on long runs and can save much time. Basically, the system consists of a tank and air pump with accessories. An electrician can vacuum or blow a line or tape in any size conduit from 1/2 through 4 inches, or even 5- and 6-inch conduits with optional accessories.

After the fish tape is inserted in the raceway system, the wires must be firmly attached by some approved means. On short runs where only a few conductors are involved, all that is necessary is to strip the insulation from the ends of the wires, to bend these ends around the hook in the fish tape, and to tape them securely in place. Where several wires are to be pulled together over long and difficult (one with several bends) conduit runs, the wires should be staggered and the fish tape securely taped at the point of attachment so that the overall diameter is not increased any more than is absolutely necessary. Staggering is done by attaching one wire to the fish tape and then attaching the second wire a short distance behind this to the bare copper conductor of the first wire. The third wire is attached to the second wire, and so forth.

Basket grips (Figure 8–8) are available in many sizes, for almost any size and combination of conductors. They are designed to hold the conductors firmly to the fish tape and can save the electrical workers much of the time and trouble that would be required to tape wires.

In all but very short runs, the wires should be lubricated with a good quality of wire lubricant prior to attempting and also during the pull. Some of this lubricant should also be applied to the inside of the conduit itself.

When wire from coils in boxes is pulled, the wire should be pulled from the center of the coil in order to prevent excessive tangling and kinking. If several conductors are to be fed into a single raceway system, they should be kept parallel and straight, free from kinks and bends. Wires that are permitted to cross each other during the pulling operation form a bulge that makes pulling difficult, especially around bends.

Wire dispensers are great aids in keeping the conductors straight and facilitating the pulling of conductors. One type is designed to accept up to 6 boxes of wire from size No. 16 AWG to No. 10 AWG. A feeding eye in the top of the dispenser helps to eliminate kinks and tangles. Large 10-inch semipneumatic tires make it convenient to move the entire apparatus (including the boxes of wire) to various locations.

Figure 8–8   BASKET GRIPS ARE AVAILABLE IN MANY SIZES, FOR ALMOST ANY SIZE OR COMBINATION OF CONDUCTORS.

Other types of dispensers are designed for spools or wires. These compact wire dispensers handle up to 10 spools of wire from No. 22 AWG to No. 10 AWG. The spindles are designed to permit each payout of the wire with automatic braking. Guide heads gather and straighten the wires for easy feeding, eliminate kinks or tangles. One type even has a guide head that rotates 360° and can be raised or lowered approximately 30° for pulling wires into a conduit system from any angle.

8–8   WIRE CONNECTIONS

Anyone who is involved in the installation of electrical systems should have a good knowledge of wire connectors and splicing since it is necessary to make numerous electrical joints during the course of any electrical installation. Splices and connections that are properly made will often last as long as the insulation on the wire itself. Poorly made connections will always be a source of trouble, i.e., the joints will overheat under load and will cause a higher resistance in the circuit than there should be. The basic requirements for a good electrical connection include:

1   It should be mechanically and electrically secure.

2   It should be insulated as well as or better than the existing insulation on the conductors.

3   These characteristics should last as long as the conductor is in service.

There are many different types of electrical joints for different purposes, and the selection of the proper type for a given application depends to a great extent on how and where the splice or connection is used. Electrical joints in this day and age are normally made with solderless pressure connectors in order to save time.

Wire connections are used to connect a wire to an electrical device such as a receptacle, wall switch, or pump control switch.

Electricians normally encounter wiring devices with screw terminals more often than any other type, and the simple "eye" connection is the one to use for such terminals. To make the "eye" in the wire, strip and clean approximately 1 to 1-1/2 inches of insulation from the end of the wire. With a pair of long-nose pliers, make a slight bend in the wire near the insulation at an angle of approximately 45°. Continue by bending the wire (above the first bend) in the opposite direction and at different points to form a circle, as shown in Figure 8–9. The "eye" may then be placed under the screw head so that the direction of the second bend in the wire is the same as the direciton in which the screw will be tightened since this will cause the "eye" to close tightly around the screw threads. If the "eye" is reversed, the "eye" will open and be loose around the threads (see Figure 8–10).

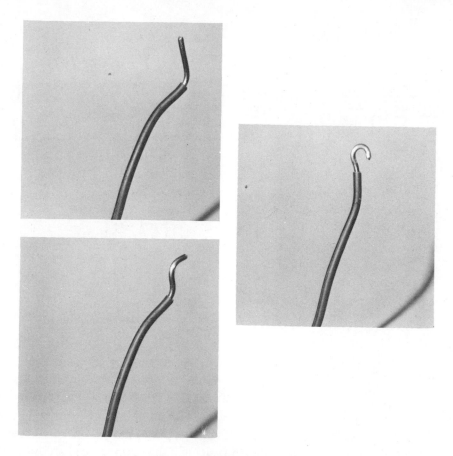

Figure 8–9   PROCEDURES FOR MAKING AN "EYE" IN A WIRE.

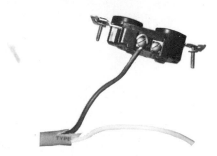

Figure 8–10   IMPROPER WAY OF USING "EYE" UNDER A TERMINAL SCREW.

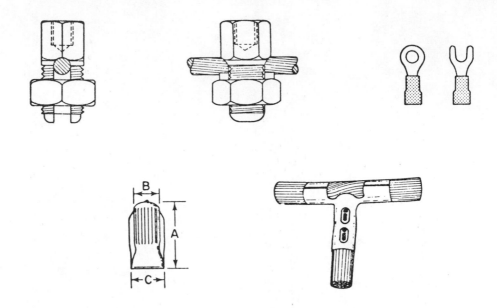

Figure 8–11   VARIOUS TYPES OF WIRE CONNECTORS.

Two other types of wire connectors are shown in Figure 8–11. Both are used for connecting wires to screw terminals. To install on a wire, strip and clean a length of insulation from the wire exactly the length of the slot on the connector. Insert the bare end of the wire in the open slot and then crimp the slot down tightly against the wire with a pair of pliers or a crimping tool. With the wire now secure in the connector, the connector's eye or ears are inserted under the screw head of the terminal, and the screw is tightened for a second electrical connection.

Other types of wire connectors are available, and direction of their use will normally be found on the carton in which they are shipped. Some require special tools for proper use, but most may be used for electrical splices and connections with conventional hand tools.

QUESTIONS

Answer the following questions by filling in the blanks.

1   The maximum number of No. 10 AWG conductors permitted in a 4-inch × 1-1/2 inch octagonal outlet box without clamps, wiring devices, and grounding wires is _____.

2   Outlet boxes must be securely and rigidly fastened to the surface upon which they are mounted or embedded in _____ or _____.

3   One type of area in which type NM cable may not be used is in _____.

4   Type NM and AC cable must be supported within _____ inches of each outlet box or other termination point.

5   Clean threads are required for a continuous _____.

6   The NEC states that a 90° bend in rigid conduit must have a minimum radius of _____ times the inside diameter of the conduit.

7   Smaller sizes of rigid conduit are normally bent with a _____ or _____.

8   _____-type benders are always used to bend EMT.

9   Surface metal raceways must not be used in areas that are subjected to _____ vapors, in _____, nor in any _____ location.

10   In all but very short runs, wires should be well _____ prior to attempting the pull.

9

FEEDER INSTALLATIONS

Electric power is delivered to panelboard locations by feeder conductors that extend from the main service entrance equipment to the branch circuit panelboards. The feeder conductors must be of sufficient size to meet the requirement of NEC Article 220 and are calculated roughly on a minimum basis of allowing for the connected lighting load or a certain number of watts per square foot for lighting, plus the power-equipment or appliance-load requirements. In some cases, a demand factor is allowed for the connection power load.

On larger installations, distribution centers, to which point large feeder conductors are run from the main service equipment, are established. Subfeeders are run from the distribution center to the various lighting and power panelboards.

When the service entrance voltage is higher than that required for the branch circuit wiring, step-down transformers are used near the panelboard locations to reduce the voltage of the conductors entering the panelboards. The usual voltages in such cases are 480/277 volts for the feeders and 120/240 volts for the branch-circuit panelboards.

In some instances, especially on industrial applications, higher voltage feeders, such as 2300 volts, are provided. The principle is the same as described in the previous paragraph. The difference is in the conductor insulation and type of transformer and switching equipment.

Feeder raceways from the main distribution center to the branch circuit panelboards or heavy electrical equipment may be installed under concrete slabs; exposed on walls, ceilings, or beams and trusses; or concealed in walls or in spaces forming the finished ceilings. Vertical feeder runs are often installed in pipe shafts provided in the building structure. The main points for the electrician's consideration are:

1   The proper wiring method must be used.

2   Raceways or cables must be secured as required by the NEC.

3   Conductors must be of the proper size and have the proper insulation for the use intended.

4   Terminations must be performed in a proper manner, using correct connectors and other materials.

9–1   CALCULATING FEEDER SIZE

Calculating the proper size feeder for a system involves determining the total amperage of the load and then selecting a conductor size from NEC tables with an ampacity sufficient for the load. For individual pieces of equipment where one set of feeder conductors will supply one piece of equipment, the total load (full-load amperes) of the equipment may be found on the nameplate attached to the equipment or from manufacturer's literature, in case the equipment is not present during the installation of the feeder raceway.

The type of load also plays a role in selecting the feeder conductors. For example, conductors supplying a single motor must have an ampacity of not less than 125 percent of the motor, full-load, current rating. Where a feeder supplies continuous loads or any combination of continuous and noncontinuous loads, neither the ampere rating of the overcurrent device nor the ampacity of the feeder conductors should be less than the noncontinuous load plus 125 percent of the continuous load.

Demand Factors

The demand factors listed in Table 9–1 apply to that portion of the total branch-circuit load computed for general illumination of a building or premises. However, it must not be applied in determining the number of branch circuits for general illumination.

Table 9–1   LIGHTING LOAD FEEDER DEMAND FACTORS

Type of Occupancy	Portion of lighting load to which demand factor applies (wattage)	Demand factor (percent)
Dwelling units	First 3000 or less	100
	Next 3001 to 120,000	35
	Remainder over 120,000	25
Hospitals	First 50,000 or less	40
	Remainder over 50,000	20

Type of Occupancy	Portion of lighting load to which demand factor applies (wattage)	Demand factor (percent)
Hotels and motels—including apartment houses without provision for cooking by tenants	First 20,000 or less Next 20,001 to 100,000 Remainder over 100,000	50 40 30
Warehouses (storage)	First 12,500 or less Remainder over 12,500	100 50
All others	Total Wattage	100

Tables 9–2—9–10 are for other electrical loads, where a demand factor applies.

Table 9–2   DEMAND FACTORS FOR NONDWELLING RECEPTACLE LOADS

Portion of receptacle load to which demand factor applies (wattage)	Demand factor (percent)
First 10 kW or less	100
Remainder over 10 kW	50

Table 9–3   DEMAND FACTORS FOR HOUSEHOLD ELECTRIC CLOTHES DRYERS

Number of dryers	Demand factor (percent)
1	100
2	100
3	100
4	100
5	80
6	70
7	65
8	60
9	55
10	50
11-13	45
14-19	40
20-24	35
25-29	32.5
30-34	30
35-39	27.5
40 & over	25

Table 9–4 DEMAND LOADS FOR HOUSEHOLD ELECTRIC RANGES, WALL-MOUNTED OVENS, COUNTER-MOUNTED COOKING UNITS, AND OTHER HOUSEHOLD COOKING APPLIANCES OVER 1-3/4 kW RATING

| Number of appliances | Maximum demand | | |
	Column a (not over 12 kw rating) (kw)	Column b (less than 3-1/2 kw rating) (percent)	Column c (3-1/2 kw to 8-3/4 kw rating) (percent)
1	8	80	80
2	11	75	65
3	14	70	55
4	17	66	50
5	20	62	45
6	21	59	43
7	22	56	40
8	23	53	36
9	24	51	35
10	25	49	34
11	26	47	32
12	27	45	32
13	28	43	32
14	29	41	32
15	30	40	32
16	31	39	28
17	32	38	28
18	33	37	28
19	34	36	28
20	35	35	28
21	36	34	26
22	37	33	26
23	38	32	26
24	39	31	26
25	40	30	26
26-30	15 kW plus 1 kW	30	24
31-40	for each range	30	22
41-50	25kW plus 3/4	30	20
51-60	kW for each	30	18
61 & over	range	30	16

Table 9–5 FEEDER DEMAND FACTORS FOR KITCHEN EQUIPMENT—OTHER THAN DWELLING UNITS(s)

Number of units of equipment	Demand factors (percent)
1	100
2	100
3	90
4	80
5	70
6 & over	65

Table 9–6 OPTIONAL CALCULATION FOR DWELLING UNIT

Load (in kw or kva)	Demand factor (percent)
Air conditioning and cooling, including heat pump compressors	100
Central electric space heating	65
Less than four separately controlled electric space heating units	65
First 10 kW of all other load	100
Remainder of other load	40

Table 9–7 OPTIONAL CALCULATION—DEMAND FACTORS FOR THREE OR MORE MULTIFAMILY DWELLING UNITS

Demand factor (percent)	Number of dwelling units
3-5	45
6-7	44
8-10	43
11	42
12-13	41
14-15	40
16-17	39
18-20	38
21	37
22-23	36
24-25	35
26-27	34
28-30	33
31	32

Demand factor (percent)	Number of dwelling units
32-33	31
34-36	30
37-38	29
39-42	28
43-45	27
46-50	26
51-55	25
56-61	24
62 & over	23

Table 9–8   OPTIONAL METHOD—DEMAND FACTORS FOR FEEDERS AND SERVICE-ENTRANCE CONDUCTORS FOR SCHOOLS

Connected load (watts per square foot)	Demand factors (percent)
Connected load up to and including 3, plus	100
Connected load over 3 and including 20, plus	75
Connected load over 20	25

Table 9–9   METHOD FOR COMPUTING FARM LOADS FOR OTHER THAN DEWLLING UNIT

Ampere load at 230 volts	Demand factor (percent)
Loads expected to operate without diversity, but not less than 125 percent full-load current of the largest motor and not less than the first 60 amperes of load	100
Next 60 amperes of all other loads	50
Remainder of other load	25

Table 9–10   METHOD FOR COMPUTING TOTAL FARM LOAD

Individual loads computed in accordance with table 9–9	Demand factor (percent)
Largest Load	100
Second Largest Load	75
Third Largest Load	65
Remaining Loads	50

9–2   VOLTAGE DROP

In all electrical wiring, the conductors should be sized so that the voltage drop never exceeds 3 percent for power, heating, lighting loads, or combinations thereof. Furthermore, the maximum total voltage drop for conductors for feeders and branch circuits should never exceed 5 percent overall.

The voltage drop in any two-wire, single-phase circuit consisting of mostly resistance-type loads with negligible inductance may be found by the following equation:

$$VD = \frac{2K \times L \times I}{CM}$$

where, VD = drop in circuit voltage, r = resistance per foot of conductor, i = current in the circuit, CM = area of conductor in circular mils, and K = resistivity of conductor materials (11 for copper and 18 for aluminum).

With this equation, the voltage drop in a circuit consisting of No. 10 AWG wire, 50 feet in length, and carrying a load of 20 amperes would be:

$$VD = \frac{2(\text{in equation}) \times 50 \ (\text{length}) \times 20 \ (\text{current}) \times 11 \ (K)}{10,380 \ (\text{area in CM of No. 10 wire})}$$

$$= \quad 2.11 \text{ volts}$$

Figures 9–1 through 9–3 give examples of voltage drop tables; the proper use of such tables can save the electrician much valuable time when he is working with voltage drops. For example, to use the table in Figure 9–1, first determine the total demand load in amperes; assume this to be 130 amperes. Further assume that the length of the feeder run is 300 feet. Find 130 amperes in the very left-hand column and follow the row corresponding to this 130-ampere figure across until the column under 300 feet is reached. Read the wire size as 4/0; this is the proper size aluminum conductor to use on a 240-volt circuit in order to keep the voltage drop within 3 percent.

9–3   DETERMINING SIZE OF CONDUIT FOR NUMBER AND SIZES OF WIRE

Tables 3A, 3B, and 3C of the NEC (Chapter 9) give the maximum number of conductors that may be installed in trade sizes of conduit. These numbers apply only to complete raceway systems and are not intended to apply to short sections of conduit used to protect exposed wiring from physical damage.

When equipment grounding conductors are installed in the raceway system, they must be included in the calculations of the conduit fill. The actual dimensions of the equipment grounding conductor (insulated or bare) must be used in the calculation.

Figure 9–1 ALUMINUM CONDUCTORS, 230–240 VOLTS, SINGLE PHASE, 3 PERCENT VOLTAGE DROP.

Compare size shown below with size shown to left of double line. Use the larger size.

Load in Amp	In Cable, Conduit, Earth — Types R, T, TW	In Cable, Conduit, Earth — Types RH, RHW, THW	Overhead in Air* Bare or Covered Conductors — Single	Overhead in Air* — Triplex	75	100	125	150	175	200	250	300	350	400	450	500	550	600	650	700	750	800	900	1000
5	12	12	10	10	12	12	12	12	12	12	12	12	10	10	10	8	8	8	8	8	6	6	6	6
7	12	12	10	10	12	12	12	12	12	12	12	10	10	8	8	8	8	8	6	6	6	6	6	4
10	12	12	10	10	12	12	12	10	10	10	8	8	6	6	6	6	6	6	4	4	4	4	4	2
15	12	12	10	10	12	10	10	8	8	8	6	6	4	4	4	4	4	4	2	2	2	2	2	1
20	10	10	10	10	10	10	8	8	8	6	6	4	4	4	2	2	2	2	2	2	1	1	1	0
25	10	10	10	10	10	8	8	6	6	6	4	4	2	2	2	2	1	1	1	0	0	0	0	00
30	8	8	10	10	8	8	6	6	6	4	4	2	2	2	1	1	0	0	0	0	00	00	00	000
35	6	8	10	10	8	8	6	6	4	4	2	2	2	1	1	0	0	00	00	00	00	000	000	000
40	6	8	10	10	8	6	6	4	4	4	2	2	1	1	0	0	00	00	000	000	000	000	4/0	4/0
45	4	6	10	10	8	6	4	4	4	2	2	1	1	0	00	00	000	000	000	4/0	4/0	4/0	4/0	250
50	4	6	8	8	6	6	4	4	2	2	1	0	0	00	00	000	000	000	4/0	4/0	4/0	4/0	250	250
60	2	4	6	6	6	4	4	2	2	2	1	0	00	00	000	000	4/0	4/0	4/0	4/0	250	250	300	300
70	2	2(a)	6	4	6	4	2	2	2	1	0	00	000	000	4/0	4/0	4/0	250	250	250	300	300	350	350
80	1	2(a)	6	4	4	4	2	2	1	0	00	000	000	4/0	4/0	4/0	250	250	300	300	300	350	350	400
90	0	2(a)	4	2	4	4	2	1	1	0	000	000	4/0	4/0	4/0	250	250	300	300	350	350	350	400	500
100	0	1(a)	4	2	4	2	2	1	0	00	000	4/0	4/0	4/0	250	250	300	300	350	350	400	400	500	500
115	00	0(a)	2	1	000	000	0	0	0	00	000	000	4/0	250	300	300	350	350	400	400	500	500	500	600
130	000	00(a)	2	0	000	000	00	00	00	000	000	4/0	250	300	300	350	400	400	500	500	500	600	600	700
150	4/0	000(a)	1	00	4/0	4/0	000	000	000	4/0	4/0	250	300	300	400	400	500	500	500	600	600	600	700	750
175	300	4/0(a)	0	000	250	250	4/0	4/0	4/0	4/0	250	300	300	350	500	500	500	600	600	700	700	700	800	900
200	350	250	00	4/0	350	300	250	250	250	250	300	350	400	400	500	500	600	600	700	700	750	800	900	1M
225	400	300	000		000	000	000	4/0	4/0	250	300	350	350	400	500	500	600	600	700	700	800	800	900	1M
250	500	350	000		000	000	000	4/0	4/0	250	350	350	400	500	500	600	600	700	700	800	800	900	1M	
275	600	500	4/0		4/0	4/0	250	250	300	300	350	400	500	500	600	600	700	700	800	900	900	1M		
300	700	500	250		250	250	300	300	300	300	400	500	500	600	600	700	750	800	900	1M	1M			
325	800	600	300		300	300	300	300	350	350	400	500	500	500	600	600	700	700	800	900	900	1M		
350	900	700	300		300	300	300	350	350	350	500	500	600	600	700	700	800	800	900	1M	1M			
375	1M	700	350		350	350	350	350	350	400	500	500	600	700	700	750	900	900	1M					
400		900	350		350	350	350	350	400	400	500	600	600	700	750	800	900	1M						

Minimum Allowable Size of Conductor

Length of Run in Feet

| Motor Hp | Full Load Amp | Minimum Allowable Size of Conductor | | | Length of Run in Feet | | | | | | | | | | | | | |
| | | In Cable, Conduit, Earth | | Overhead in Air* | Compare size shown below with size shown to left of double line. Use the larger size. | | | | | | | | | | | | | |
		Types R, T, TW	Types RH, RHW, THW	Bare & Covered Conductors	100	150	200	300	400	500	600	700	800	900	1000	1200	1500	2000
7½	11	14	14	10	14	14	14	12	12	10	10	8	8	8	8	6	6	4
10	14	12	12	10	12	12	12	12	10	10	8	8	8	6	6	6	4	4
15	20	10	10	10	10	10	10	10	8	8	6	6	6	6	4	4	3	2
20	26	8	8	10	10	10	8	8	8	6	6	6	4	4	4	3	2	1
25	32	8	8	10	10	10	8	8	6	6	4	4	4	4	3	2	1	0
30	39	6	6	10	10	8	8	6	6	4	4	4	3	3	2	1	0	00
40	52	4	6	8	8	6	6	6	4	4	3	3	2	2	1	0	00	000
50	63	2	4	6	6	6	6	4	4	3	2	2	1	1	0	00	000	4/0
60	75	2	3	6	6	4	4	4	3	2	2	1	1	1	00	000	000	250
75	93	0	1	4	4	4	4	2	2	1	1	0	0	0	000	000	4/0	300
100	123	000	00	2	2	2	2	1	1	0	0	00	000	000	4/0	4/0	300	400
125	155	4/0	000	1	1	1	1	0	0	00	00	000	4/0	4/0	250	300	350	500
150	180	300	4/0	0	0	0	0	0	0	00	000	4/0	4/0	250	300	350	400	600

Figure 9–2 COPPER CONDUCTORS, THREE-PHASE MOTORS, 440 VOLTS, 3 PERCENT VOLTAGE DROP.

Minimum Allowable Size of Conductor — In Cable, Conduit, Earth; Overhead in Air*

Length of Run in Feet — Compare size shown below with size shown to left of double line. Use the larger size.

Load in Amp	Types R, T, TW	Types RH, RHW, THW	Bare or Covered Conductors	50	60	75	100	125	150	175	200	225	250	275	300	350	400	450	500	550	600
5	14	14	10	14	14	14	14	14	12	12	12	10	10	10	10	8	8	8	8	6	6
7	14	14	10	14	14	14	12	12	12	10	10	8	8	8	8	8	6	6	6	6	6
10	14	14	10	14	14	12	12	10	10	8	8	8	8	6	6	6	6	6	4	4	4
15	14	14	10	12	12	12	10	8	8	6	6	6	6	6	4	4	4	3	3	3	2
20	12	12	10	12	10	10	8	8	6	6	6	6	4	4	4	3	3	2	2	2	1
25	10	10	10	10	10	8	8	6	6	6	4	4	4	4	3	3	2	2	1	1	0
30	10	10	10	10	8	8	6	6	4	4	4	3	3	2	2	1	1	1	0	0	00
35	8	8	10	10	8	8	6	6	4	4	4	3	3	2	2	1	1	0	0	00	00
40	8	8	10	8	8	6	6	4	4	4	3	2	2	2	1	1	0	00	00	000	000
45	6	8	10	8	8	6	6	4	4	3	2	2	1	1	1	0	00	00	000	000	000
50	6	6	10	8	6	6	4	4	3	3	2	2	1	1	0	0	00	000	000	000	4/0
60	4	6	8	6	6	6	4	3	2	2	1	1	0	0	00	00	000	000	000	4/0	4/0
70	4	4	8	6	6	4	4	3	2	1	1	0	0	00	00	000	000	4/0	4/0	250	250
80	2	4	6	6	4	4	3	2	1	0	0	00	00	000	000	4/0	4/0	250	300	300	300
90	2	3	6	6	4	4	2	2	1	00	00	000	000	4/0	4/0	250	300	300	350	300	350
100	1	3	6	4	4	3	2	1	0	00	00	000	000	4/0	4/0	250	300	300	300	350	350
115	0	2	4	4	4	2	1	0	0	0	000	000	4/0	4/0	250	300	350	350	350	400	500
130	00	1	4	4	2	2	1	0	00	000	4/0	4/0	250	250	300	350	350	400	500	500	500
150	000	0	2	2	2	2	0	00	000	4/0	4/0	250	250	300	350	350	400	500	600	600	600
175	4/0	00	1	2	2	1	0	00	000	4/0	250	300	350	350	400	500	500	600	700	700	700
200	250	000	1	1	1	0	00	000	4/0	250	300	300	350	350	500	500	600	600	700	700	700

Figure 9-3   COPPER CONDUCTORS, 115–120 VOLTS, SINGLE PHASE, 3 PERCENT VOLTAGE DROP.

When conduit nipples having a maximum length of 24 inches are installed between boxes, cabinets, and similar enclosures, the nipple may be filled to 60 percent of its total cross-sectional area. The conduit tables are shown in Figure 9–4, 9–5, and 9–6.

Example 9–1

The example presented in Section 9–2 required size 4/0 conductors to maintain the proper limits of voltage drop. Assume three 4/0 conductors are used in a conduit. What size conduit must be used?

Solution

Table 3A of the NEC gives a 2-inch conduit as the proper trade size for three 4/0 conductors with TW, THW, RHW, etc. insulation.

9–4   DETERMINING THE TYPE OF WIRE INSULATION BEST SUITED FOR THE AREA USED

Chapter 3 of the NEC gives requirements for wiring methods and materials for electrical systems in which voltages do not exceed 600 volts. Article 310 of the NEC covers general requirements for conductors and their type designations, insulations, markings, mechanical strengths, ampacity ratings, and uses. Table 310–13 of the NEC (reproduced in Figure 9–7) gives conductor applications and insulations for use at 600 volts or less. Article 310, Sections 310–30 through 310–61 cover use of conductors on systems over 600 volts.

9–5   INSTALLING FEEDERS

As soon as the building lines have been established on any given project, the electricians should install all necessary conduit sleeves before the footings are poured. This is also the time to orient plans for installing all heavy electrical equipment in relation to fixed and identical building points. When these points have been established, the electricians can positively fix the location of main distribution service equipment, panelboards, transformers, and other equipment requiring large conduit sizes and conductors. Once it has been established that feeder conduits will not interfere with the general building construction, these raceways should be installed.

Number of Conductors	1	2	3	4	Over 4
All conductor types except lead-covered (new or rewiring)	53	31	40	40	40
Lead-covered conductors	55	30	40	38	35

Figure 9–4   PERCENT OF CROSS SECTION OF CONDUIT AND TUBING FOR CONDUCTORS.

Conduit Trade Size (Inches)	½					¾					1					1¼					1½					2				
Wire Types	18	16	14	12	10	18	16	14	12	10	18	16	14	12	10	18	16	14	12	10	18	16	14	12	10	18	16	14	12	10
PTF, PTFF, PGFF, PGF, PFF, PF, PAF, PAFF, ZF, ZFF	23	18	14			40	31	24			65	50	39			115	90	70			157	122	95			257	200	156		
TFFN, TFN	19	15				34	26				55	43				97	76				132	104				216	169			
SF-1	16					29					47					83					114					186				
SFF-1, FFH-1	15					26					43					76					104					169				
CF	13	10	8	4	3	23	18	14	7	6	38	30	23	12	9	66	53	40	21	16	91	72	55	29	22	149	118	90	48	37
TF	11	10				20	18				32	30				57	53				79	72				129	118			
RFH-1	11					20					32					57					79					129				
TFF	11	10				20	17				32	27				56	49				77	66				126	109			
AF	11	9	7	4	3	19	16	12	7	5	31	26	20	11	8	55	46	36	19	15	75	63	49	27	20	123	104	81	44	34
SFF-2	9	7	6			16	12	10			27	20	17			47	36	30			65	49	42			106	81	68		
SF-2	9	8	6			16	14	11			27	23	18			47	40	32			65	55	43			106	90	71		
FFH-2	9	7				15	12				25	19				44	34				60	46				99	75			
RFH-2	7	5				12	10				20	16				36	28				49	38				80	62			
KF-1, KFF-1, KF-2, KFF-2	36	32	22	14	9	64	55	39	25	17	103	89	63	41	28	182	158	111	73	49	248	216	152	100	67	406	353	248	163	110

Figure 9–5   MAXIMUM NUMBER OF FIXTURE WIRES IN TRADE SIZES OF CONDUIT OR TUBING.

Type Letters	Conductor Size AWG, MCM	½	¾	1	1¼	1½	2	2½	3	3½	4	4½	5	6
TW, T, RUH, RUW, XHHW (14 thru 8)	14	9	15	25	44	60	99	142						
	12	7	12	19	35	47	78	111	171					
	10	5	9	15	26	36	60	85	131	176				
	8	2	4	7	12	17	28	40	62	84	108			
RHW and RHH (without outer covering), THW	14	6	10	16	29	40	65	93	143	192				
	12	4	8	13	24	32	53	76	117	157				
	10	4	6	11	19	26	43	61	95	127	163			
	8	1	3	5	10	13	22	32	49	66	85	106	133	
TW, T, THW, RUH (6 thru 2), RUW (6 thru 2)	6	1	2	4	7	10	16	23	36	48	62	78	97	141
	4	1	1	3	5	7	12	17	27	36	47	58	73	106
	3	1	1	2	4	6	10	15	23	31	40	50	63	91
	2	1	1	2	4	5	9	13	20	27	34	43	54	78
	1		1	1	3	4	6	9	14	19	25	31	39	57
FEPB (6 thru 2), RHW and RHH (without outer covering)	0		1	1	2	3	5	8	12	16	21	27	33	49
	00		1	1	1	3	5	7	10	14	18	23	29	41
	000		1	1	1	2	4	6	9	12	15	19	24	35
	0000			1	1	1	3	5	7	10	13	16	20	29
	250			1	1	1	2	4	6	8	10	13	16	23
	300			1	1	1	2	3	5	7	9	11	14	20
	350					1	1	3	4	6	8	10	12	18
	400				1	1	1	3	4	5	7	9	11	16
	500				1	1	1	1	3	4	6	7	9	14
	600					1	1	1	3	4	5	6	7	11
	700					1	1	1	2	3	4	5	7	10
	750					1	1	1	2	3	4	5	6	9

Figure 9–6   MAXIMUM NUMBER OF CONDUCTORS IN TRADE SIZES OF CONDUIT AND TUBING.

Type Letters	Conductor Size AWG, MCM	½	¾	1	1¼	1½	2	2½	3	3½	4	4½	5	6
THWN,	14	13	24	39	69	94	154							
	12	10	18	29	51	70	114	164						
	10	6	11	18	32	44	73	104	160					
	8	3	5	9	16	22	36	51	79	106	136			
THHN, FEP (14 thru 2), FEPB (14 thru 8), PFA (14 thru 4/0), PFAH (14 thru 4/0), Z (14 thru 4/0), XHHW (4 thru 500MCM)	6	1	4	6	11	15	26	37	57	76	98	125	154	
	4	1	2	4	7	9	16	22	35	47	60	75	94	137
	3	1	1	3	6	8	13	19	29	39	51	64	80	116
	2	1	1	3	5	7	11	16	25	33	43	54	67	97
	1		1	1	3	5	8	12	18	25	32	40	50	72
	0		1	1	3	4	7	10	15	21	27	33	42	61
	00		1	1	2	3	6	8	13	17	22	28	35	51
	000		1	1	1	3	5	7	11	14	18	23	29	42
	0000		1	1	1	2	4	6	9	12	15	19	24	35
	250			1	1	1	3	4	7	10	12	16	20	28
	300			1	1	1	3	4	6	8	11	13	17	24
	350			1	1	1	2	3	5	7	9	12	15	21
	400				1	1	1	3	5	6	8	10	13	19
	500				1	1	1	2	4	5	7	9	11	16
	600					1	1	1	3	4	5	7	9	13
	700						1	1	1	3	4	5	6	11
	750						1	1	1	2	3	4	6	11
XHHW	6	1	3	5	9	13	21	30	47	63	81	102	128	185
	600					1	1	1	3	4	5	7	9	13
	700						1	1	1	3	4	5	6	11
	750						1	1	1	2	3	4	6	10

Figure 9–6   (Continued)

Type Letters	Conductor Size AWG, MCM	½	¾	1	1¼	1½	2	2½	3	3½	4	4½	5	6
RHW,	14	3	6	10	18	25	41	58	90	121	155			
	12	3	5	9	15	21	35	50	77	103	132			
	10	2	4	7	13	18	29	41	64	86	110	138		
	8	1	2	4	7	9	16	22	35	47	60	75	94	137
RHH (with outer covering)	6	1	1	2	5	6	11	15	24	32	41	51	64	93
	4	1	1	1	3	5	8	12	18	24	31	39	50	72
	3	1	1	1	3	4	7	10	16	22	28	35	44	63
	2		1	1	3	4	6	9	14	19	24	31	38	56
	1		1	1	1	3	5	7	11	14	18	23	29	42
	0		1	1	1	2	4	6	9	12	16	20	25	37
	00			1	1	1	3	5	8	11	14	18	22	32
	000			1	1	1	3	4	7	9	12	15	19	28
	0000				1	1	2	4	6	8	10	13	16	24
	250				1	1	1	3	5	6	8	11	13	19
	300				1	1	1	3	4	5	7	9	11	17
	350				1	1	1	2	4	5	6	8	10	15
	400					1	1	1	3	4	6	7	9	14
	500					1	1	1	3	4	5	6	8	11
	600						1	1	2	3	4	5	6	9
	700						1	1	1	3	3	4	6	8
	750						1	1	1	3	3	4	5	8

Figure 9–6   (Concluded)

Table 310-13. Conductor Application and Insulations

Trade Name	Type Letter	Max. Operating Temp.	Application Provisions	Insulation	AWG or MCM	Thickness of Insulation Mils	Outer Covering
Heat-Resistant Rubber	RH	75°C 167°F	Dry locations.	Heat-Resistant Rubber	**14-12 .... 30 10 .... 45 8-2 60		*Moisture-resistant, flame-retardant, non-metallic covering
Heat-Resistant Rubber	RHH	90°C 194°F	Dry locations.		1-4/0 .... 80 213-500 .... 95 501-1000 ......110 1001-2000 125		
Moisture and Heat-Resistant Rubber	RHW	75°C 167°F	Dry and wet locations. For over 2000 volts insulation shall be ozone-resistant.	Moisture and Heat-Resistant Rubber	14-10 .... 45 8-2 .... 60 1-4/0 .... 80 213-500 .... 95 501-1000 ......110 1001-2000 125		*Moisture-resistant, flame-retardant, non-metallic covering
Heat-Resistant Latex Rubber	RUH	75°C 167°F	Dry locations.	90% Un-milled, Grainless Rubber	14-10 .... 18 8-2 25		Moisture-resistant, flame-retardant, non-metallic covering

* Outer covering shall not be required over rubber insulations which have been specifically approved for the purpose.
** For 14-12 sizes RHH shall be 45 mils thickness insulation.
For insulated aluminum and copper-clad aluminum conductors, the minimum size shall be No. 12. See Tables 310-16 through 310-19.

Trade Name	Type Letter	Max. Operating Temp.	Application Provisions	Insulation	AWG or MCM	Thickness of Insulation Mils	Outer Covering
Moisture-Resistant Latex Rubber	RUW	60°C 140°F	Dry and wet locations.	90% Un-milled, Grainless Rubber	14-10 .... 18 8-2 25		Moisture-resistant, flame-retardant, non-metallic covering
Thermoplastic	T	60°C 140°F	Dry locations.	Flame-Retardant, Thermoplastic Compound	14-10 .... 30 8 .... 45 6-2 .... 60 1-4/0 .... 80 213-500 .... 95 501-1000 ......110 1001-2000 125		None
Moisture-Resistant Thermoplastic	TW	60°C 140°F	Dry and wet locations.	Flame-Retardant, Moisture-Resistant Thermoplastic	14-10 .... 30 8 .... 45 6-2 .... 60 1-4/0 .... 80 213-500 .... 95 501-1000 ......110 1001-2000 125		None
Heat-Resistant Thermoplastic	THHN	90°C 194°F	Dry locations.	Flame-Retardant, Heat-Resistant Thermoplastic	14-12 .... 15 10 .... 20 8-6 .... 30 4-2 .... 40 1-4/0 .... 50 250-500 .... 60 501-1000 70		Nylon jacket or equivalent

Figure 9-7 CONDUCTOR APPLICATIONS AND INSULATIONS.

Table 310-13 (Continued)

Trade Name	Type Letter	Max. Operating Temp.	Application Provisions	Insulation	AWG or MCM / Thickness of Insulation / Mils		Outer Covering
Moisture- and Heat-Resistant Thermoplastic	THW	75°C 167°F 90°C 194°F	Dry and wet locations. Special applications within electric discharge lighting equipment. Limited to 1000 open-circuit volts or less. (Size 14-8 only as permitted in Section 410-31.)	Flame-Retardant, Moisture- and Heat-Resistant Thermoplastic	14-10 .... 45 8-2 .... 60 1-4/0 .... 80 213-500 .... 95 501-1000 ......110 1001-2000 125		None
Moisture- and Heat-Resistant Thermoplastic	THWN	75°C 167°F	Dry and wet locations.	Flame-Retardant, Moisture- and Heat-Resistant Thermoplastic	14-12 .... 15 10 ....20 8-6 .... 30 4-2 .... 40 1-4/0 .... 50 250-500 .... 60 501-1000 70		Nylon jacket or equivalent
Moisture- and Heat-Resistant Cross-Linked Synthetic Polymer	XHHW	90°C 194°F 75°C 167°F	Dry locations. Wet locations.	Flame-Retardant Cross-Linked Synthetic Polymer	14-10 .... 30 8-2 .... 45 1-4/0 .... 55 213-500 .... 65 501-1000 .... 80 1001-2000 95		None
Moisture-, Heat- and Oil-Resistant Thermoplastic	MTW	60°C 140°F 90°C 194°F	Machine tool wiring in wet locations as permitted in NFPA Standard No. 79. (See Article 670.) Machine tool wiring in wet locations as permitted in NFPA Standard No. 79. (See Article 670.)	Flame-Retardant, Moisture-, Heat- and Oil-Resistant Thermoplastic	(A) (B) 22-12 .... 30 15 10 .... 30 20 8 .... 45 30 6 .... 60 30 4-2 ...... 60 40 1-4/0 .... 80 50 213-500 .. 95 60 501-1000 .110 70		(A) None (B) Nylon jacket or equivalent

For insulated aluminum and copper-clad aluminum conductors, the minimum size shall be No. 12. See Tables 310-16 through 310-19.

Trade Name	Type Letter	Max. Operating Temp.	Application Provisions	Insulation	AWG or MCM / Thickness of Insulation / Mils		Outer Covering
Perfluoro-alkoxy	PFA	90°C 194°F 200°C 392°F	Dry locations. Dry locations — special applications.	Perfluoro-alkoxy	14-10 .... 20 8-2 .... 30 1-4/0 45		None
Perfluoro-alkoxy	PFAH	250°C 482°F	Dry locations only. Only for leads within apparatus or within raceways connected to apparatus. (Nickel or nickel-coated copper only.)	Perfluoro-alkoxy	14-10 .... 20 8-2 .... 30 1-4/0 45		None
Extruded Polytetra-fluoroethylene	TFE	250°C 482°F	Dry locations only. Only for leads within apparatus or within raceways connected to apparatus, or as open wiring. (Nickel or nickel-coated copper only.)	Extruded Polytetra-fluoro-ethylene	14-10 .... 20 8-2 .... 30 1-4/0 45		None
Thermoplastic and Asbestos	TA	90°C 194°F	Switchboard wiring only.	Thermo-plastic and Asbestos	Th'pl'. Asb. 14-8 ..... 20 20 6-2 ...... 30 25 1-4/0 .....40 30		Flame-retardant, nonmetallic covering
Thermoplastic and Fibrous Outer Braid	TBS	90°C 194°F	Switchboard wiring only.	Thermo-plastic	14-10 .... 30 8 .... 45 6-2 .... 60 1-4/0 80		Flame-retardant, nonmetallic covering
Synthetic Heat-Resistant	SIS	90°C 194°F	Switchboard wiring only.	Heat-Resistant Rubber	14-10 .... 30 8 .... 45 6-2 .... 60 1-4/0 80		None

Figure 9–7   (Continued)

Table 310-13 (Continued)

Mineral Insulation (Metal Sheathed)	MI	85°C 185°F 250°C 482°F	Dry and wet locations. For special application.	Magnesium Oxide	16-10 .... 36 9-4 .... 50 3-250 55	Copper
Underground Feeder & Branch-Circuit Cable-Single Conductor. (For Type UF cable employing more than one conductor see Article 339.)	UF	60°C 140°F	See Article 339	Moisture-Resistant	14-10 .... *60 8-2 .... *80 1-4/0 .... *95	Integral with insulation
		75°C** 167°F		Moisture- and Heat-Resistant		
Underground Service-Entrance Cable-Single Conductor. (For Type USE cable employing more than one conductor see Article 338.)	USE	75°C 167°F	See Article 338	Heat- and Moisture-Resistant	12-10 .... 45 8-2 .... 60 1-4/0 .... 80 213-500 .... 95 501-1000 ......110 1001-2000 125	Moisture-resistant non-metallic covering [See 338-1 (2).]

* Includes integral jacket.
** For ampacity limitation, see 339-1(a).
  The nonmetallic covering over individual rubber-covered conductors of aluminum-sheathed cable and of lead-sheathed or multiconductor cable shall not be required to be flame retardant. For Type MC cable, see Section 334-20. For nonmetallic-sheathed cable, see Section 336-2. For Type UF cable, see Section 339-1.
  For insulated aluminum and copper-clad aluminum conductors, the minimum size shall be No. 12. See Tables 310-16 through 310-19.

Trade Name	Type Letter	Max. Operating Temp.	Application Provisions	Insulation	AWG or MCM	Thickness of Insulation	Mils	Outer Covering
Silicone-Asbestos	SA	90°C 194°F	Dry locations.	Silicone Rubber	14-10 .... 45 8-2 .... 60 1-4/0 .... 80 213-500 .... 95 501-1000 ....110 1001-2000 125			Asbestos or glass
		125°C 257°F	For special application.					
Fluorinated Ethylene Propylene	FEP or FEPB	90°C 194°F 200°C 392°F	Dry locations. Dry locations — special applications.	Fluorinated Ethylene Propylene	14-10 .... 20 8-2 30			None
				Fluorinated Ethylene Propylene	14-8 14			Glass braid
					6-2 14			Asbestos braid
Modified Fluorinated Ethylene Propylene	FEPW	75°C 90°C	Wet locations. Dry locations.	Modified Fluorinated Ethylene Propylene	14-10 .... 20 8-2 30			None
Modified Ethylene Tetrafluoro-ethylene	Z	90°C 194°F 150°C 302°F	Dry locations. Dry locations — special applications.	Modified Ethylene Tetrafluoro-ethylene	14-12 .... 15 10 .... 20 8-4 .... 25 3-1 .... 35 1/0-4/0 45			None
Modified Ethylene Tetrafluoro-ethylene	ZW	75°C 167°F 90°C 194°F 150°C 302°F	Wet locations. Dry locations. Dry locations — special applications.	Modified Ethylene Tetrafluoro-ethylene	14-10 .... 30 8-2 45			None

Figure 9–7   (Continued)

Table 310-13 (Continued)

Trade Name	Type Letter	Max. Operating Temp.	Application Provisions	Insulation	AWG or MCM / Thickness	Outer Covering
Varnished Cambric	V	85°C 185°F	Dry locations only. Smaller than No. 6 by special permission.	Varnished Cambric	14-8 .... 45 6-2 .... 60 1-4/0 .... 80 213-500 .... 95 500-1000 ....110 1001-2000 125	Nonmetallic covering or lead-sheath

Asbestos and Varnished Cambric	AVA	110°C 230°F	Dry locations only.	Impregnated Asbestos and Varnished Cambric	
					          AVA AVL
					     1st 2nd 2nd
					     Asb. VC Asb. Asb.
					14-8 (solid only ... — 30 20 25
					14-8 .... 10 30 15 25
					6-2 ..... 15 30 20 25
					1-4/0 ... 20 30 30 30
					213-500 . 25 40 40 40
					501-1000. 30 40 40 40
					1001-2000 30 50 50 50
						AVA-asbestos braid or glass

Asbestos and Varnished Cambric	AVL	110°C 230°F	Dry and wet locations.		(see above)	AVL-lead sheath
					VC Asb. 18-8 .... 30 20 6-2 .... 40 30 1-4/0 .... 40 40	Flame-retardant, cotton braid (switchboard wiring)
Asbestos and Varnished Cambric	AVB	90°C 194°F	Dry locations only.	Impregnated Asbestos and Varnished Cambric	2nd Asb. VC Asb. 14-8 .... 10 30 15 6-2 .... 15 30 20 1-4/0 .... 20 30 30 213-500 ..... 25 40 40 501-1000 .... 30 40 40 1001-2000 ... 30 50 50	Flame-retardant, cotton braid

For insulated aluminum and copper-clad aluminum conductors, the minimum size shall be No. 12. See Tables 310-16 through 310-19.

Trade Name	Type Letter	Max. Operating Temp.	Application Provisions	Insulation	AWG or MCM	Thickness of Insulation	Mils	Outer Covering
Asbestos	A	200°C 392°F	Dry locations only. Only for leads within apparatus or within raceways connected to apparatus. Limited to 300 volts.	Asbestos	14		30 40	Without asbestos braid
					12-8			
Asbestos	AA	200°C 392°F	Dry locations only. Only for leads within apparatus or within raceways connected to apparatus or as open wiring. Limited to 300 volts.	Asbestos	14 .... 30 12-8 .... 30 6-2 .... 40 1-4/0 60			With asbestos braid or glass
Asbestos	AI	125°C 257°F	Dry locations only. Only for leads within apparatus or within raceways connected to apparatus. Limited to 300 volts.	Impregnated Asbestos	14 .... 30 12-8 40			Without asbestos braid
Asbestos	AIA	125°C 257°F	Dry locations only. Only for leads within apparatus or within raceways connected to apparatus or as open wiring.	Impregnated Asbestos	Sol. Str. 14 .... 30 30 12-8 .... 30 40 6-2 .... 40 60 1-4/0 .... 60 75 213-500 .... 90 501-1000 105			With asbestos braid or glass
Paper		85°C 185°F	For underground service conductors, or by special permission.	Paper				Lead sheath

For insulated aluminum and copper-clad aluminum conductors, the minimum size shall be No. 12. See Tables 310-16 through 310-19.

Figure 9–7   (Concluded)

When parallel runs of conduit are being installed, always make certain that enough space is allowed between each run for locknuts, bushings, and couplings. The parallel runs should be installed simultaneously rather than progressively installing one raceway at a time. See Figure 9–8.

Whenever possible, the entire raceway system should be completed before any wire is pulled. If the conductors are installed before the entire raceway system is finished, it is possible that some wire may be damaged, requiring costly extra labor and materials to correct the damage.

Pull boxes and junction boxes (Figure 9–9) are provided in a raceway system to facilitate the pulling of conductors or to provide a junction point for the connections of conductors, or both. In some instances, the location and size of pull boxes is designated on the drawings or in the written specifications supplied by the architect or his engineer. However, in most cases, it is the electrician's responsibility to size these boxes correctly according to good practices or the NEC.

Regardless of whether junction or pull boxes are specified for a particular project, long runs of wires should not be made in one pull. Pull boxes, installed at convenient intervals relieve much of the strain on the wires and make the pull much easier. Since there is no set rule regarding the distance between pull boxes, the workmen will have to use good judgment concerning where they are required or would be beneficial.

Pull boxes should always be installed in a location that allows the workmen to work easily and conveniently. In an installation where the conduit run is routed up a corner of a wall and changes direction at the ceiling, a pull box that is installed too high will force the electrician to stand on a ladder when feeding or pulling wires. This situation is just an example; there are several situations to consider when locations for junction boxes are related. They must be securely fastened in place on walls or ceiling or some other adequate support.

To make certain that the raceway system provides a continuous equipment ground, all connections between sections of conduit and between the conduit and termination points (panels, wire troughs, junction boxes, etc.) must be tight. This is insured by the use of two locknuts on every termination point, even though metal bushings are used.

Where a conduit enters a building at the main distribution panel and the grounding is provided at this point, it is a good idea to install bonding bushings on the panel to insure continuity. Electrical disturbances may occur in the wiring system, both inside and out, and many contractors do not like to rely entirely on locknuts and bushings for electrical continuity at this point in the electrical system where a good ground is very important.

After the ground is made, the electrical contractor should check the system with a megger to make certain that the resistance is low enough to qualify under the requirements of the NEC. When a ground wire is to be connected to a water pipe, the area around where the ground clamp will be secured should be

Figure 9–8   PARALLEL RUNS OF CONDUIT SHOULD BE INSTALLED SIMULTANEOUSLY RATHER THAN PROGRESSIVELY INSTALLING ONE CONDUIT AT A TIME.

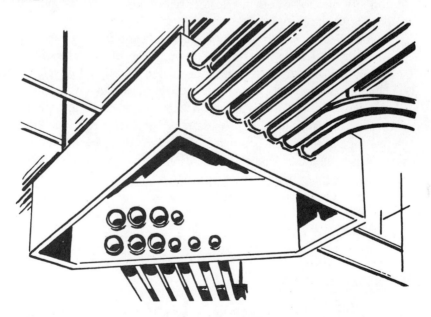

Figure 9–9   PULL BOXES AND JUNCTION BOXES ARE USED IN RACEWAY SYSTEMS TO FACILITATE THE PULLING AND CONNECTION OF CONDUCTORS.

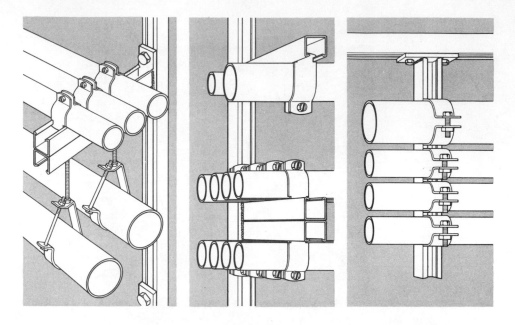

Figure 9–10   APPLICATION OF METAL SUPPORT CHANNELS AND SINGLE-BOLT PIPE STRAPS.

thoroughly cleaned of all dirt, paint, rust, and other substances that might offer resistance to the flow of electricity.

The selection of the proper hangers and supports is a very important function in roughing-in feeder raceways. Although the electrical estimator will include some hangers and supports in his bid, he will usually allow a lump sum figure for such items and have the exact selection of these devices to men in the field.

Raceways installed exposed on masonry surfaces are usually held in place with one- or two-hole straps, the straps being securely anchored with lead anchors, toggle bolts, and similar means. Two or more parallel raceway systems surface mounted are usually installed on special metal channels and are held in place by matching pipe straps. This method seems to be the most popular since it is the fastest way to install banks of conduit and gives a very neat appearance. Figure 9–10 shows several applications of metal support channels and single-bolt pipe straps.

9–6   CONDUIT BENDS

Although there are not as many conduit bends in service and feeder conduit systems as there are in branch circuit wiring, there are still a few installations in which factory ells take care of all the required bends. Electricians are required to make several complicated offsets, saddles, radius bends, and other bends requiring both skill and calculations.

The NEC specifies the minimum radius of conduit bends, which varies from six to eight times the inside diameter of the conduit. Table 9–11 gives the minimum radius of conduit bends for 1-1/4-inch through 6-inch conduit sizes.

Table 9–11

Size of conduit (inches)	Minimum radius of bend (inches)
1-1/4	8
1-1/2	10
2	12
2-1/2	15
3	18
3-1/2	21
4	24
5	30
6	36

9–7 PULLING THE WIRE

Whenever possible, feeder conductors should be pulled directly from the reels without prehandling them. This can be accomplished by ordering conductors precut to the required length and wound on the reels at the factory in three or four conductor pulls. This requires extremely close checking of the drawings and adequate allowances for lengths of conductors in the raceway system. However, the savings in labor cost on the job will normally more than offset the cost of such a coordinating effort. As an extra precaution against errors in calculating the lengths of conductors involved, it is good practice to measure physically all runs with a fish tape before starting the cable pull and checking the totals against the totals indicated on the reels.

It is usually best to pull cables downward rather than upward, to avoid having to pull the total weight of the cable at the final stages of the pull and to avoid the possibility of injury to workers should the conductors break loose from the pulling cable on long vertical pulls. However, if heavy feeder cables are to be pulled downward for any appreciable length, the cable reel must be filled with some type of brake so that the reel will not turn too fast because of the dead weight of the vertical cable runs. Also, the inner ends of the cables (at the reel core) must be anchored so that the ends do not get away from the pulling gang when the end of the pull is approached.

Conductors that are to be installed downward should be fed off the top of the reel; where conductors are to be fed upward, the best method is to feed from the bottom of the reel. This procedure eliminates sharp kinks or bends in the conductors.

When the conduit is installed through a pull box, it should enter and leave the box in a manner that allows the greatest possible sweep to the conductors. Large conductors, especially, are difficult to bend, and the electrician can simplify the feeding of these conductors from one conduit to another by proper planning. Hydraulic and ratchet cable-bending tools are now available that will quickly and easily bend 350-MCM to 100-MCM cable.

Consideration must also be given to conductor supports in vertical raceways. Article 300–19 of the NEC requires that one cable support be provided at the top of the vertical raceway or as close to the top as practical as well as a support for each additional interval of spacing, as specified in the NEC. Such devices can save much time in handling large sizes of conductors since they are designed to make one-shot bends up to 90° and then to automatically unload the cable.

Compact portable power pullers that allow one man to pull a considerable amount of cable in a raceway system are available. Most of these tools weigh less than 50 pounds and are designed for set up in less than 5 minutes for use on 1-1/4-inch through 2-1/2-inch conduits. A font switch on the puller leaves both hands free during the pull. When cable is pulled through manholes or on a cable tray system, hook-type cable sheaves (Figure 9–11) should be used to change the direction of the cable under pull, to suspend the cable on long spans, or to keep the cables bunched together for overhead pulls.

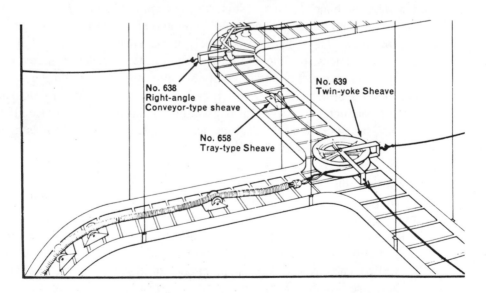

Figure 9–11   HOOK-TYPE CABLE SHEAVES SHOULD BE USED TO CHANGE THE DIRECTION OF A CABLE UNDER PULLED IN MANHOLES.

9–8   TERMINATIONS

The termination of the conductors at points of connection must be done with care in order to prevent expenditure of unnecessary time repairing the results of poor connections. In making such connections, three factors are involved: tightness of the connection, area of contact, and insulation at the point of termination.

Looseness of connections or limited area of contact will cause heating at the point of connection that will reduce the efficiency of the circuit because of the resulting higher resistance and will actually damage the conductor insulation and the equipment. Proper size and type of connecting lugs, securely tightened, must be used.

QUESTIONS

Answer the following questions by filling in the blanks.

1   Article _____ of the NEC covers most provisions for designing and installing feeder circuits.

2   Conductors feeding a single motor must have an ampacity of not less than _____ percent of the motor full-load current.

3   In hospitals, the first 50,000 watts of lighting load can be calculated by using a _____-percent demand factor.

4   In all electrical installations, the wiring should be sized so that the voltage drop never exceed _____ percent for power, heating, and lighting loads.

5   In the voltage drop equation, $VD = \dfrac{2K \times L \times I}{CM}$ CM stands for _____.

6   Four 500-MCM conductors with THW insulation require a _____-inch trade size conduit.

7   Table _____ of the NEC gives conductor applications and insulation.

8   When installing parallel runs of conduit, always make certain that enough space is allowed between each run for _____, _____, and _____.

9   One reason for installing pull boxes in a raceway system is to facilitate the _____ of conductors.

10   _____ locknuts should be used on every conduit termination point to insure grounding continuity.

10

MOTOR CIRCUITS
AND FEEDERS

The principal means of changing electrical energy into mechanical energy is the electric motor, which ranges in size from small, fractional horsepower motors that operate on 120 or 240 volts to the very large, high-voltage, synchronous and induction motors that are used for industrial applications. Article 430 of the NEC covers design, application, and installation standards of motor circuits and motor control connections, including disconnects, starters, running overload protection, and short-circuit and ground-fault protection. The basic elements required in most motor circuits include:

1   Motor branch-circuit conductors rated at least 125 percent of the motor's full-load (rated) current (Article 430, Part B).

2   Branch-circuit protective devices (both overload and short circuit) to protect the motor and the circuit feeders from short circuits and ground faults (Article 430, Part D).

3   Motor controller to start, stop, and control power to the motor in a way suited to the load and the supply system (Article 430, Part G).

4   Running overload (overcurrent) protection to open (clear) the circuit on overloads up to and including the motor's rated locked-rotor current (Article 430, Part C).

5   Disconnecting means to disconnect both motor and controller from all ungrounded supply conductors (Article 430, Part H).

6   Control circuit for manual or automatic operation of the motor starter (Article 430, Part F).

10–1   SIZING CIRCUITS FOR MOTORS

There are three factors to consider when the size of motor feeder conductors are determined: the length of the motor feeder and motor branch-circuit run, the motor full-load (rated) current, and the permissible voltage drop (Article 430, Part B). The length of feeder circuit may be determined by scaling the working drawings or by actually measuring at the job site. The full-load current drawn by a particular motor should appear on its nameplate. If this information is not available, the approximate full-load current can be found in Tables 430–148, 430–149, and 430–150 of the NEC. First determine the horsepower of the motor in question, and then look under the appropriate column for the voltage, motor type, and phases used to find the full-load current in amperes.

Example 10–1

Find the full-load currents of the following motors:

a   5-hp, 115-V a.c., single-phase motor

b   3-hp, 230-V a.c., single-phase motor

c   3-hp, 208-V a.c., single-phase motor

d   30-hp, 230-V a.c., three-phase, squirrel-cage motor

e   30-hp, 208-V a.c., three-phase, squirrel-cage motor

f   30-hp, 220-V a.c., three-phase, synchronous motor at 0.8 P.F.

Solution

a   56 amp (directly from Table 430–148)

b   17 amp (directly from Table 430–148)

c   1.1 × 17 amp = 18.7 A (10 percent increase, as noted in Table 430–148)

d   80 amp (directly from Table 430–150)

e   1.1 × 80 amp = 88 A (10 percent increase, as noted in Table 430–150)

f   1.25 × 65 amp = 81.25 A (25 percent increase, as noted in Table 430–150)

Before conductors for motor feeders are sized, the length of the run and the allowable voltage drop must be taken into consideration. The NEC states that "the size of the conductors for feeders should be such that the voltage drop for the load as computed by Section 220–4 would not be more than 3 percent for power, heating of lighting loads or combinations thereof. . . ." Therefore, the maximum allowable voltage drop for a 120-volt circuit is (120 × 0.03 =) 3.6 volts; similarly, for 240-volt circuits, the volt drop is (240 × 0.03 =) 7.2 volts. With this information, the conductor size may be determined by Equation 10–1:

$$\text{Conductor size} = \frac{2\,K \times \text{length of conductor} \times \text{current}}{\text{allowable voltage drop}}$$

or

$$A = 2K \times L \times I/V \qquad\qquad (10\text{--}1)$$

where; A is the conductor size in circular mils (CM); I is the current amperes (amp); L is the length of conductors, i.e., the length of the circuit one way, in feet (ft); V is the voltage drop in actual volts not percentage (V); K is the resistivity in ohms-circular mil per foot (Ω-CM/ft); K equals 11 for copper and 18 for aluminum conductors.

The circular mil values that correspond to the various wire gauge sizes (AWG) are shown in Figure 10–1. They are also available in Table 8, Chapter 9 of the NEC.

Size AWG MCM	Area Cir. Mils	Concentric Lay Stranded Conductors		Bare Conductors		D.C. Resistance Ohms/M Ft. At 25°C. 77°F.		
		No. Wires	Diam. Each Wire Inches	Diam. Inches	*Area Sq. Inches	Copper		Aluminum
						Bare Cond.	Tin'd. Cond.	
18	1620	Solid	.0403	.0403	.0013	6.51	6.79	10.7
16	2580	Solid	.0508	.0508	.0020	4.10	4.26	6.72
14	4110	Solid	.0641	.0641	.0032	2.57	2.68	4.22
12	6530	Solid	.0808	.0808	.0051	1.62	1.68	2.66
10	10380	Solid	.1019	.1019	.0081	1.018	1.06	1.67
8	16510	Solid	.1285	.1285	.0130	.6404	.659	1.05
6	26240	7	.0612	.184	.027	.410	.427	.674
4	41740	7	.0772	.232	.042	.259	.269	.424
3	52620	7	.0867	.260	.053	.205	.213	.336
2	66360	7	.0974	.292	.067	.162	.169	.266
1	83690	19	.0664	.332	.087	.129	.134	.211
0	105600	19	.0745	.372	.109	.102	.106	.168
00	133100	19	.0837	.418	.137	.0811	.0843	.133
000	167800	19	.0940	.470	.173	.0642	.0668	.105
0000	211600	19	.1055	.528	.219	.0509	.0525	.0836
250	250000	37	.0822	.575	.260	.0431	.0449	.0708
300	300000	37	.0900	.630	.312	.0360	.0374	.0590
350	350000	37	.0973	.681	.364	.0308	.0320	.0505
400	400000	37	.1040	.728	.416	.0270	.0278	.0442
500	500000	37	.1162	.813	.519	.0216	.0222	.0354
600	600000	61	.0992	.893	.626	.0180	.0187	.0295
700	700000	61	.1071	.964	.730	.0154	.0159	.0253
750	750000	61	.1109	.998	.782	.0144	.0148	.0236
800	800000	61	.1145	1.030	.833	.0135	.0139	.0221
900	900000	61	.1215	1.090	.933	.0120	.0123	.0197
1000	1000000	61	.1280	1.150	1.039	.0108	.0111	.0177
1250	1250000	91	.1172	1.289	1.305	.00863	.00888	.0142
1500	1500000	91	.1284	1.410	1.561	.00719	.00740	.0118
1750	1750000	127	.1174	1.526	1.829	.00616	.00634	.0101
2000	2000000	127	.1255	1.630	2.087	.00539	.00555	.00885

Figure 10–1   PROPERTIES OF ELECTRICAL CONDUCTORS.

Example 10–2

A 120-V, two-wire, lighting branch circuit runs a length (one way) of 100 feet from its overcurrent protection of 30 amperes to its first lighting outlet. When all the lights are ON, a current of 25 amperes is drawn. Calculate the conductor size in CM and AWG using:

a   Copper conductors

b   Aluminum conductors

Solution

a $A = 2 \times K \times L \times I/V$

       $= 2 \times 12 \times 100 \times 25/3.6 = 16{,}666.7$ Cmil

   $A =$ No. 6 AWG (Figure 10–1)

b $A = 16{,}667$ Cmil $\dfrac{18}{25} = 25{,}000$ Cmil

      $=$ No. 6 AWG aluminum conductors (Figure 10–1).

In three-phase motor circuits, the voltage drop between any two outside conductors (nongrounded conductors) equals 0.866 times the drop determined. Therefore, the following equation may be used to determine the size of conductors for a three-phase motor circuit or feeder:

$$\text{Conductor size (Cmil)} = \frac{2K \times \text{current} \times \text{length (one way)}}{\text{allowable voltage drop}} \times 0.886$$

or

$$A = 2K \times L \times I \times 0.886/V \qquad \textbf{(10–2)}$$

where all terms have been defined previously.

Temperature Considerations

Temperature is also an important factor in the selection of conductors for motor feeders. The ambient temperature of all areas in which the conductors pass should be taken into consideration when the conductor insulation is selected since excess heat increases the resistance of the wire, which in turn increases the voltage drop for any given load.

Conductor insulation is rated from 60°C to 200°C—giving a wide selection from which to choose. If the routing of the motor circuit will pass only in areas where the ambient temperature is below 86°F, conductors with type TW or THW insulation are satisfactory. However, if the conductors, for example, run through a boiler room where the temperatures may be as high as 190°F, then conductors with type THHN, RHH, AVA, or similar insulation should be used.

10–2 WIRE SIZE FOR MOTOR FEEDERS

To illustrate the method of sizing conductors for motor feeders, assume that a 5-hp, single-phase motor is to be fed. The motor is located 300 feet from a 120/240-volt, main service panelboard and the full-load ampere rating is 26.5 amperes at 240 volts. Copper conductors ($K = 11$) will be used for the circuit. Substituting into Equation 10–1, we have:

$$A = 2 \times K \times L \times I/V$$
$$= \frac{2(11) \times 300 \times 26.5}{7.2} = 24,292 \text{ Cmil}$$

From the calculation, the minimum size of the conductor is 24,292 Cmils. In the table in Figure 10–1, 26,240 circular mils is the closest larger standard wire size to the answer; this is No. 6 AWG.

It is also desirable not to have the voltage of the circuit drop below 200 volts on *starting*, This allows a 40-volt drop if the initial voltage is 240 volts. If the starting current is three times the running current, the substitutions in the equation become:

$$\text{Wire Size} = \frac{2 \times 11 \times (3 \times 26.5) \times 300}{40} = 13,118 \text{ Cmil}$$

Since this answer is less than the 24,292 CM required for the running current, the original selection of a No. 6 AWG conductor (26,240 CM) is quite satisfactory for *both* the starting and running current.

10–3 OVERCURRENT PROTECTION

Circuits for individual motors normally require two types of overcurrent protection: branch circuit or feeder short-circuit protection and motor overload protection. The former is provided to protect the motor system feeders and to prevent motor overload currents that might prove harmful. The latter is intended to protect the motor circuit components, such as the switches, controllers, overload relays, and motor windings.

The current rating of a fuse or circuit breaker for motor branch circuit and short-circuit protection may be as follows:

1   For short-circuit protection, the rating of a *nontime-delay* fuse may be sized at 300 percent or less of the motor full-load ampere rating for ordinary motors.

2   For overload protection, the rating of dual-element, *time-delay* fuses should never exceed 225 percent of the motor full-load ampere rating.

3   For short-circuit protection, the setting of a trip circuit breaker (*without time delay*) may be increased to 700 percent of the motor full-load current.

4   For overload protection, the rating of a *time-delay* circuit breaker may be increased to 250 percent of the motor full-load rating.

Running overcurrent protection needs to be provided only on motors used for continuous duty with a rating over 1 hp. This overcurrent protection may be in the form of an external overcurrent device actuated by the motor running current and set to open at 125 percent to 140 percent of the motor full-load current, depending on the motor service factor and its temperature rise.

Example 10–3

A 20-hp, three-phase, 240-V, squirrel-cage induction motor (SCIM) draws a full-load current of 54 amperes from its supply. Use the information in Section 10–3 to calculate:

a   The maximum overload protection, using a dual-element time-delay fuse.

b   The minimum short-circuit protection, using a nontime-delay fuse.

Solution

a   Maximum overload protection = $2.25 \times I_{fl}$ = 122 amperes. The closest standard fuse that is *below* the maximum is a 100-ampere fuse (see NEC 240–23), time-delay type.

b   Minimum short-circuit protection, using a nontime-delay fuse = $3 \times I_{fl}$ = 162 amperes. The closest standard fuse that is *above* the minimum is a 200-ampere fuse (Sec NEC 240–23).

10–4   TERMINATION OF CONDUCTORS

When motors are provided with terminal housing, the housing must be of metal and of substantial construction. When these terminal housings enclose wire-to-wire connections, they should have minimum dimensions and usable volumes in accordance with Table 10–1.

Table 10–1   TERMINAL HOUSING DIMENSIONS FOR MOTORS

Horsepower	Cover opening, minimum dimension (inches)	Usable volume, minimum (cubic inches)
1 and smaller	1-5/8	7-1/2
1-1/2 through 3	1-3/4	12
5 through 7-1/2	2	16
10 through 15	2-1/4	22-1/2
20 through 25	2-7/8	33
30 through 40	3	44
50 through 60	3-1/2	72-1/2
75 through 100	3-1/2	100
125 through 150	6	216

The vibration caused by motors during operation makes the *flexible* metal conduit one of the most popular means of terminating conductors on the last 18 inches or so of a conventional rigid raceway system (rigid conduit, EMT, etc.). In general, the flexible metal conduit should be of same size (diameter) as the conventional raceway; that is, if 3/4″ rigid conduit, for example, is used to feed the motor from its controller to the last few feet of the run, the flexible metal conduit should also be of 3/4″ diameter.

When flexible metal conduit is installed, it should be secured by approved means at intervals not exceeding 4-1/2 feet and within 12 inches of each side of every outlet box, fitting, or motor terminal box. Where more flexibility is needed, lengths of not more than 3 feet may be used at terminals.

In most cases, it is a good idea to pull a *separate* grounding conductor through flexible metal conduit to insure a good ground. The NEC allows the conduit itself to be used as a ground, provided that the conduit used for grounding is 6 feet or less, that it is terminated in fittings approved for the purpose, and that the circuit conductors contained in the conduit are protected by overcurrent devices rated at 20 amperes or less.

Where the motors are used in damp areas or in areas exposed to the weather, a liquidtight, flexible metal conduit may be used for the last few feet of raceway termination to a motor housing. This type of raceway system should not be used where it will be subjected to physical damage or where any combination of ambient and/or conductor temperature will produce an operating temperature in excess of that for which the material is approved. Liquidtight, flexible metal conduits may also be used for grounding in the 1-1/4-inch and smaller sizes if the length is 6 feet or less and if it is terminated in fittings approved for the purpose. Other installation requirements are the same as for standard, flexible metal conduits.

Once the raceway system is complete and the conductors are pulled into the motor terminal housing, the conductors must be secured to the motor leads in an approved manner. For single-voltage motors that have two or three leads, *crimp-type* (solderless) connectors are normally employed to secure the circuit conductors to the motor leads, and the connectors are then taped (insulated). On dual-voltage motors on which the leads may have to be changed, the motors leads (usually nine wires on three-phase motors) and the circuit conductors are fitted with crimp-on eye connectors and are then secured to each other by a bolt and nut. The joints are taped in a conventional manner.

10–5   MOTOR CONTROLS

Devices used to "control" motors and motor circuits vary from simple toggle switches to complex systems, using components such as relays, timers, and magnetic motor starters. Some of the functions of motor controllers are starting, stopping, reversing, running, speed control, safety of operator, and protection from damage.

Figures 10–2—10–9 show many ways in which full voltage and reduced voltage starters can be used in practical, motor-circuit applications. It should become apparent from these diagrams that a multitude of other circuit combinations can be readily incorporated. The symbols shown in Figure 10–2 represent in diagram form the devices and components used in the circuits to follow. Familiarity with these symbols will help you to acquire a better understanding of how line and wiring diagrams show circuit operation.

Figure 10–3 shows a line diagram and a wiring diagram of toggle-operated, single-pole switch with a thermal overload relay. This type of control can be used where only one line needs to be interrupted (as on a 120-volt motor, 1 hp or less).

Figure 10–4 shows the connection of a toggle-operated, double-pole switch with thermal over-load relay and with a "run-Auto" selector switch and a pilot light mounted in the same enclosure. The pilot light indicates when the motor is running. This type of circuit is often used with two-wire pilot devices such as thermostats.

Diagrams of a three-pole switch with a mechanically linked pushbutton operator (without overload relay) are shown in Figure 10–5. A double-pole switch with a mechanically linked pushbutton operator and thermal over-load relay for single-phase motors is shown in Figure 10–6. Both of these starters of the full-voltage type.

Connection details of a three-phase, reversing, push-button-operated starter are shown in Figure 10–7. The switches are mechanically interlocked so that the STOP button must be depressed before directions are changed. The connections shown are for three-phase motors; for single-phase motors, the connection on lines L3 and T3 is omitted.

A two-speed, pushbutton-operated starter with a thermal overload relay for each phase is shown in Figure 10–8. Connector "B" is provided on motor starters that have two-coil, overload protection. The mechanical interlock between starters allows speed transfer to be accomplished without the STOP button being pushed. This type of starter is designed for two-speed, constant horsepower, separate winding motors.

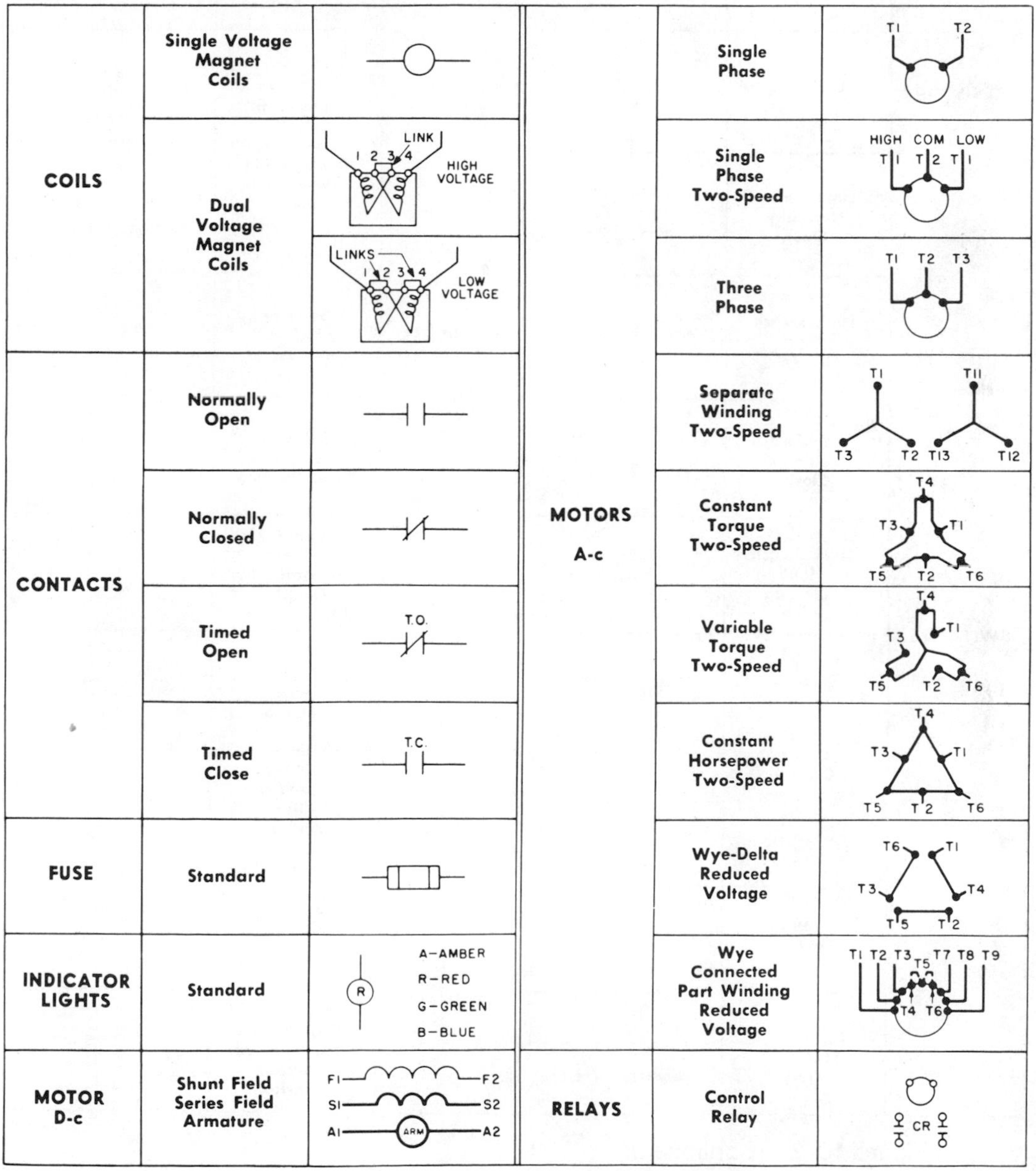

COILS	Single Voltage Magnet Coils	⊸◯⊸		Single Phase	T1 T2
	Dual Voltage Magnet Coils	LINK 1 2 3 4 HIGH VOLTAGE / LINKS 1 2 3 4 LOW VOLTAGE		Single Phase Two-Speed	HIGH COM LOW T1 T2 T1
				Three Phase	T1 T2 T3
CONTACTS	Normally Open	—⊣ ⊢—	**MOTORS** **A-c**	Separate Winding Two-Speed	T1 T11 / T3 T2 T13 T12
	Normally Closed	—⊣/⊢—		Constant Torque Two-Speed	T4 / T3 T1 / T5 T2 T6
	Timed Open	T.O. —⊣/⊢—		Variable Torque Two-Speed	T4 / T3 T1 / T5 T2 T6
	Timed Close	T.C. —⊣ ⊢—		Constant Horsepower Two-Speed	T4 / T3 T1 / T5 T2 T6
FUSE	Standard	—⊏▭⊐—		Wye-Delta Reduced Voltage	T6 T1 / T3 T4 / T5 T2
INDICATOR LIGHTS	Standard	Ⓡ A—AMBER R—RED G—GREEN B—BLUE		Wye Connected Part Winding Reduced Voltage	T1 T2 T3 T5 T7 T8 T9 / T4 T6
MOTOR D-c	Shunt Field Series Field Armature	F1 ⌇⌇⌇ F2 / S1 ⌇⌇⌇ S2 / A1 —(ARM)— A2	**RELAYS**	Control Relay	CR

Figure 10–2   WIRING AND LINE DIAGRAM SYMBOLS.

RELAYS	Thermal Overload	RESET		Pushbutton Standard Duty	NORMALLY CLOSED
	Time Delay	W/INST INTER-LOCK TC T.O. TR TC TO		Pushbutton Heavy Duty Oil Tight	
RESISTOR	Standard	R1	**SWITCHES**	"Roto-Push" Selector Pushbutton Two-Position Selector Ring	JOG RUN
SWITCHES	Limit Switch Normally Open				
	Limit Switch Held Open			Selector Switch Two-Position	JOG RUN
	Limit Switch Normally Closed			Selector Switch Three-Position	OFF RUN AUTO
	Limit Switch Held Closed			Selector Switch Mechanically Coupled Three-Position	HIGH OFF LOW
	Local Cover Control	START		Toggle Operator Manual	
		STOP		Two-Wire Pilot Devices	
	Plugging Switch		**TRANS-FORMERS**	Low Voltage Control Transformer	HI H2 X1 X2
	Pushbutton Standard Duty	NORMALLY OPEN		Auto-Transformer For Reduced Voltage Starting	% 50 65 80 100 0 % 50 65 80 100 0

Figure 10–2   (Continued)

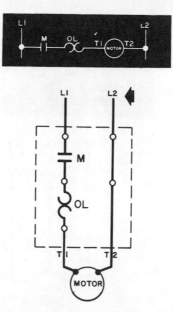

Figure 10–3 DIAGRAM OF A SINGLE-POLE SWITCH WITH A THERMAL OVERLOAD RELAY.

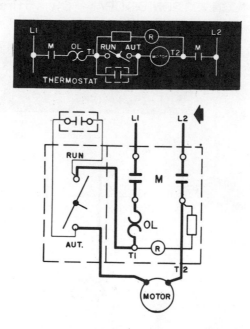

Figure 10–4 CONNECTION FOR A TOGGLE-OPERATED, DOUBLE-POLE SWITCH WITH THERMAL OVERLOAD RELAY AND A RUN-AUTO SELECTOR SWITCH WITH PILOT LIGHT.

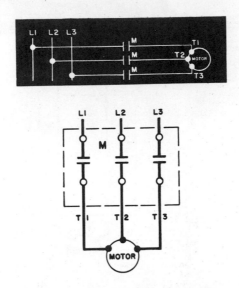

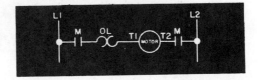

Figure 10-5   CONNECTION DIAGRAM FOR A THREE-POLE SWITCH WITH MECHANICALLY LINKED, PUSHBUTTON OPERATOR.

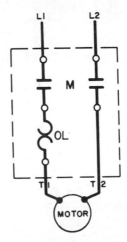

Figure 10-6   DOUBLE-POLE SWITCH WITH MECHANICALLY LINKED PUSHBUTTON OPERATOR WITH THERMAL OVERLOAD RELAY.

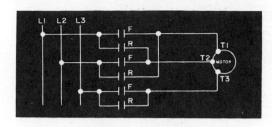

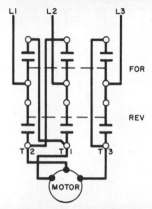

Figure 10–7 CONNECTION DETAILS OF A THREE-PHASE, REVERSING, PUSHBUTTON-OPERATED MOTOR STARTER.

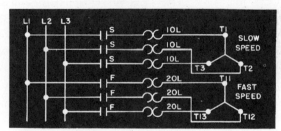

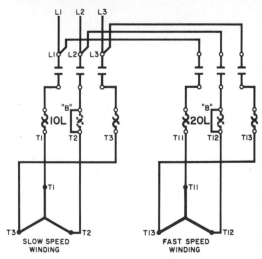

Figure 10–8 A TWO-SPEED, PUSHBUTTON-OPERATED STARTER WITH THERMAL OVERLOAD RELAY FOR EACH PHASE.

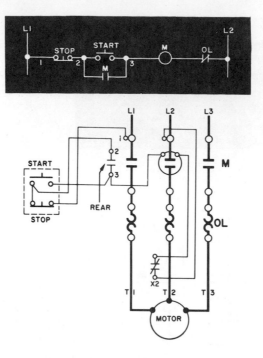

Figure 10–9 A THREE-POLE, MOTOR STARTER WITH THREE-COIL, THER-MAL OVERLOAD RELAYS AND STANDARD, THREE-WIRE, PUSHBUTTON STATION.

A three-pole, RESET ONLY motor starter wtih three-coil, thermal overload relay and standard three-wire START-STOP pushbutton station is shown in Figure 10–9. A diagram of a single-phase circuit is shown in Figure 10–10; note the insert that gives connections for dual-voltage motor coils.

The three-pole, RESET ONLY combination starter with a fusible, motor circuit switch and three-coil, thermal overload relay is shown in Figure 10–11. The starter is connected to a three-phase motor and to a standard three-wire START-STOP pushbutton station.

The wiring arrangement in Figure 10–12 finds application where a single motor is operated from several remote locations. This diagram also suggests other circuit possibilities, including multiple emergency STOP stations. A typical application is the machine operator's working in a location that is not within easy reach of the main pushbutton station.

To operate three motor starters from a *single* pushbutton location, connections such as the ones in Figure 10–13 can be used. The starters are wired so that all three will automatically be disconnected from the line if a maintained overload occurs on any one. This is accomplished by wiring the holding circuit of each starter through the auxiliary contacts of one of the other two. Since the control circuit is common among the starters, incoming power lines to all three starters must be opened by a disconnect proceding each of the starters in order to disconnect completely the starters from the line.

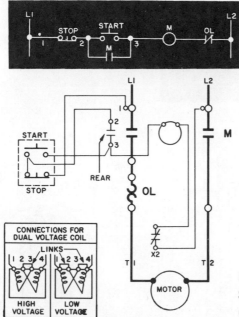

Figure 10–10   DIAGRAM FOR A SINGLE-PHASE MOTOR CONTROL CIRCUIT.

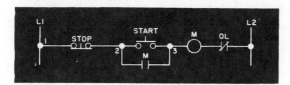

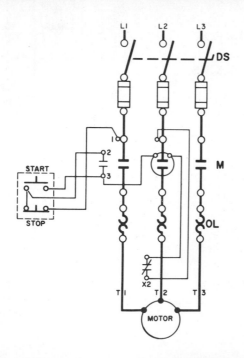

Figure 10–11   A THREE-POLE COMBINATION STARTER WITH FUSIBLE MOTOR CIRCUIT SWITCH.

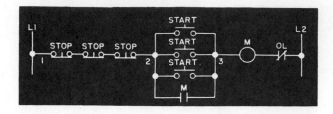

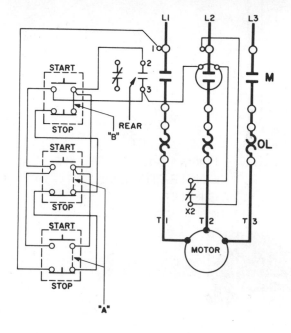

Figure 10–12 CONNECTION DIAGRAM FOR
OPERATING A SINGLE MOTOR FROM
SEVERAL DIFFERENT REMOTE LOCATIONS.

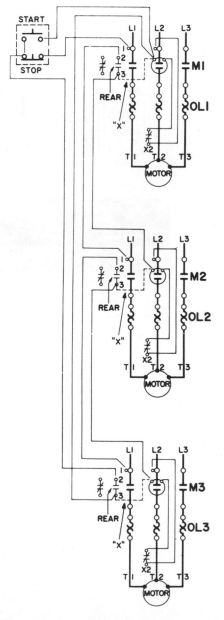

Figure 10–13 CONNECTION DIAGRAM FOR
OPERATING THREE MOTOR STARTERS FROM
A SINGLE PUSHBUTTON LOCATION.

Jogging circuits, as shown in Figure 10–14, are primarily used when machines must be operated momentarily for inching, such as in a machine tool set up for maintenance. The jog circuit energizing the starter only when the JOG button is depressed, thereby giving the machine operator instantaneous control of the motor drive. In the circuit under consideration, when the JOG button is depressed, the control relay is bypassed, and the main contractor coil is energized solely through the JOG button; when the JOG button is released, the contractor coil releases immediately. Pushing the START button closes the control relay, and it is held in by its own N.O. (normally open) contacts. The main contractor coil is in turn closed by another set of N.O. contacts on the control relay and is held on. (See Figure 10–15.)

A common single-pole, double-throw, float switch circuit is shown in Figure 10–16: the type used in sump pump and tank-filling applications. The circuit shown is connected for tank operation, but the float switch can be connected as indicated by the dotted lines in the drawing for sump pump operation. In both cases, the float within the sump pump or the tank controls the opening and closing of the contacts in the switch, which in turn controls the motor starter.

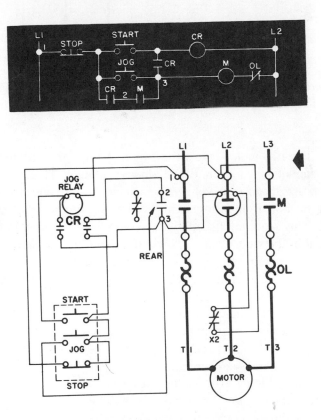

Figure 10–14   WIRING DIAGRAM FOR A TYPICAL MOTOR CONTROL CIR-CUIT UTILIZING JOGGING FEATURE.

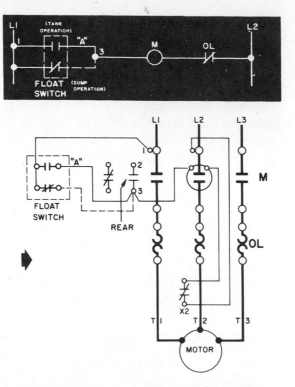

Figure 10–15 CONNECTION DETAILS FOR A COMMON SINGLE-POLE, DOUBLE-THROW, FLOAT SWITCH CIRCUIT.

The two motors shown in Figure 10–17 are arranged in such a manner that one cannot be started until the other is running. This type of control is necessary when one machine feeds a second such as in a conveyor system. To prevent the first machine from piling up material on the second, the second machine is started first. This is accomplished by interconnecting the pushbutton stations. The control circuit of the second starter is wired through the auxiliary contacts of the first; this prevents it from starting until after the first starter is energized. A timer can be connected between the starters so that the second motor will run a short time after the first motor has stopped.

Reduced-voltage starters are used primarily in applications where incoming power lines are not capable of handling the inrush current drawn by the motor during starting and in applications where mechanical stress on the driven machine may cause damage. There are several types of reduced-voltage motor starters: the diagram in Figure 10–18 illustrates a typical resistance-start type. This is called the *primary resistor* type, and its components include two contractors, a resistor bank, a three-coil, thermal overload relay, and a time-delay relay. When the START button is depressed, the ''M'' contactor and the time-delay relay (TR) are energized and the motor winding it connected at reduced voltage to the incoming

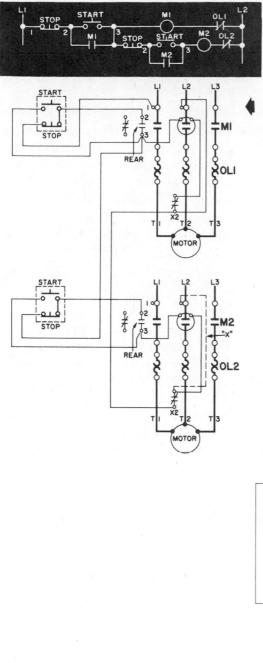

Figure 10–16 TWO PUSHBUTTONS ARRANGED SO THAT ONE MOTOR CANNOT BE STARTED UNTIL THE OTHER IS RUNNING.

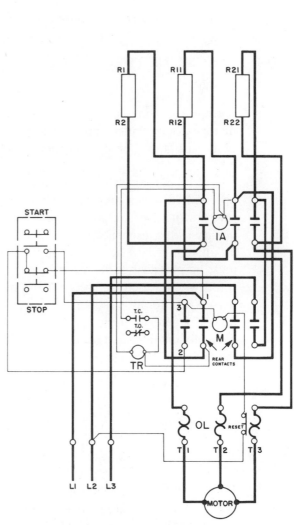

Figure 10–17 PRIMARY RESISTOR-TYPE, REDUCED-VOLTAGE, MOTOR STARTER.

power lines (primary circuit) through the series-connected resistor bank. After the time-delay relay has timed out, the timed close (T.C.) contacts close, and the "1A" contactor is energized—shorting out each of the three resistors in the resistor bank. The motor winding is then automatically switched to full voltage. The motor is stopped by depressing the STOP button, which drops out both contactors and the time-delay relay.

The autotransformer type of reduced-voltage starter operates on much the same principle as the primary resistor type. However, in the former, a transformer is used in place of the resistor for reducing the line voltage during starting. The circuit shown in Figure 10–19 is typical of those in use today. When the START button is depressed, contactors "S" and "M" and the time-delay relay are energized, applying power through the windings of the autotransformer to the motor. When the time-delay relay times out, the timed open (T.O.) contacts open and the T.C. contacts close, the "S" contactor drops out, and the "R" contactor is energized, switching the motor to full line voltage.

10–6   SPECIAL NEC REQUIREMENTS

A switch used as a disconnecting means for a motor circuit must have a horsepower rating not less than that of the motor it controls. For example, a 5-hp motor must be started and stopped by a control device that has a nameplate rating of at least 5 hp. Certain exceptions to this rule are covered by Section 430 of the NEC.

If a magnetic switch is used as a motor controller, it must have a manually operated, disconnecting switch ahead of it. The switch may be a conventional safety switch, or, in some cases, the overcurrent device (circuit breaker) may qualify as the disconnecting means. If the switch is more than 50 feet away from the motor, it must be in sight of the motor controller; otherwise, another switch must be provided in the circuit.

Switches used as disconnecting means for motor circuits should be rated in horsepower for all motors in excess of 2 hp. A switch should carry a rating of at least 125 percent of the full-load, nameplate, current rating of the motor and be manually operable in a readily accessible location. It must indicate whether it is in the open (OFF) or closed (ON) position. When opened, the switch must disconnect both the controller and the motor from all ungrounded supply conductors.

QUESTIONS

Answer the following questions by filling in the blanks.

1   The section in the NEC that deals with motors is Article _____.

2   If the full-load current of a motor cannot be obtained from the nameplate, Tables 430– _____, 430–_____, and 430–_____ may be consulted.

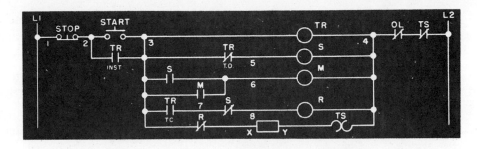

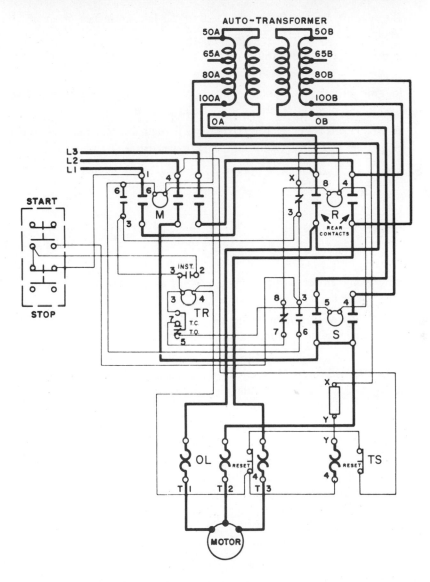

Figure 10–18 AUTOTRANSFORMER-TYPE, REDUCED-VOLTAGE, MOTOR STARTER.

3   The _____ temperature of all areas in which conductors pass should be given consideration when the type of conductors to use for motor feeders are selected.

4   Circuits used for motor feeders require _____ protection and _____ protection.

5   The rating of a nontime-delay fuse for a motor circuit may be sized at _____ percent or less of the motor full-load current.

6   Running overcurrent protection needs to be provided only on continuous duty motors over _____ hp.

7   The type of conduit frequently used on the last 18 inches or so of a raceway system feeding a motor is _____.

8   Some of the functions of a motor control are _____, _____, and _____.

9   Jogging circuits on motor controls are primarily used where machines must be _____ or _____.

10   When controls are interlocked so that one particular motor cannot start before another, the motors are more than likely furnishing power for a _____ system.

11   The _____ _____ is one type of reduced-voltage starter.

PROBLEMS

1   How many amperes does a 3-hp, 230-V a.c., single-phase motor draw? Use Table 430–148 of the NEC. _____

2   With the answer found in Problem 1 and the data presented in this chapter, what size wire will be required to feed the motor properly at a distance of 130 feet from its overcurrent protection. _____

3   What size overcurrent protection will be required? What is the maximum size time-delay fuse permitted by the NEC that can be used on this circuit? _____

11

SERVICE EQUIPMENT

All buildings, areas, and equipment that utilize an electrical system require one or more electric services—usually supplied by the local utility company. This service is defined by the NEC as "the conductors and equipment for delivering energy from the electrical supply system to the wiring system of the premises served."

A review of electric generation and transmission will help the reader to understand the purpose of an electric service. Figure 11–1 shows the basic sections of an electric transmission system. Heat energy is produced by coal in a boiler, which produces steam to drive a steam engine; this engine provides mechanical energy to drive an electric generator; electric energy from the generator is routed through a transformer by an electric conductor to raise the voltage for long-distance transmission lines; near the point of usage, another transformer is used to reduce the voltage to a safe level for use in homes, factories, and commercial buildings.

Figure 11–2 gives a more detailed description of the electric service at the point of usage. The high-voltage lines terminate on a power pole near the building that is served. A transformer is mounted on the pole to reduce the voltage to a usable level (120/240 volts in this case). The overhead conductors between the last utility company pole and the point of their connection to the building (or other support used for the purpose) are called the *service drop* (Article 100). All components between the point of termination of the overhead service drop and the building main-disconnecting device (switch), such as the service cap and the service-entrance cable, are known as the *service entrance*; the utility company's

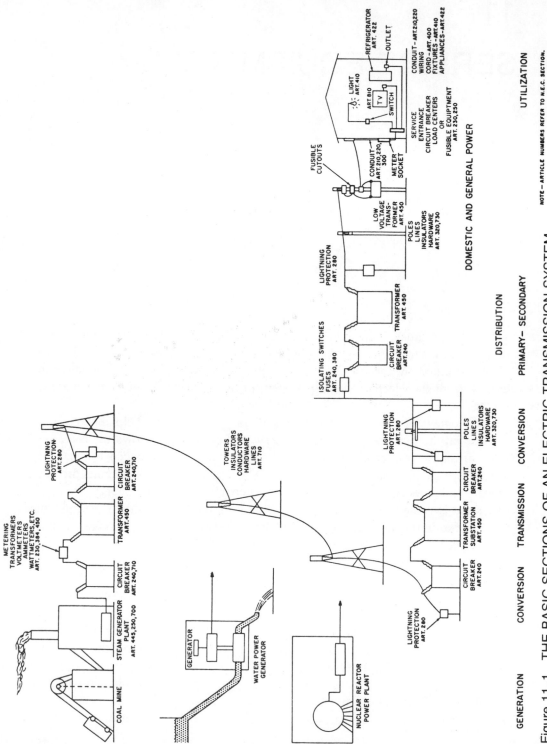

GENERATION **CONVERSION** **TRANSMISSION** **CONVERSION** **DISTRIBUTION** **PRIMARY— SECONDARY** **UTILIZATION**

DOMESTIC AND GENERAL POWER

NOTE— ARTICLE NUMBERS REFER TO N.E.C. SECTION.

Figure 11–1 THE BASIC SECTIONS OF AN ELECTRIC TRANSMISSION SYSTEM.

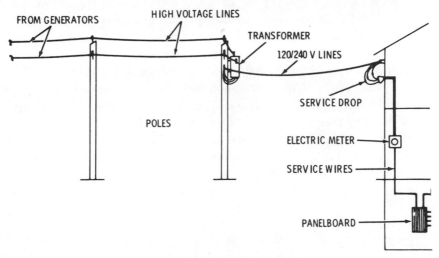

Figure 11–2   PARTS OF AN ELECTRIC SERVICE.

metering equipment is not included in this category. The wires or conductors used in this section of the electrical service are called the *service-entrance conductors*.

Service-entrance equipment provides overcurrent protection to the feeder and service conductors as well as a means of disconnecting the feeder from energized service conductors. The electric meter, which is a part of the service-entrance equipment, provides a means of measuring the amount of energy used.

11–1   SERVICE-ENTRANCE CALCULATIONS

The service-entrance conductors and equipment must have adequate ampacity in order to conduct safely the current for the loads supplied without causing a temperature rise, which could damage the insulation on the conductors. The following methods describe how to calculate the ampacity for a number of different occupancies.

Residences

In general, a residence served with electric power can be supplied through only one set of service-entrance conductors (Article 230–2). The service conductors must have adequate mechanical strength and cannot be smaller than No. 8 AWG copper or No. 6 AWG aluminum (Article 230–23). The NEC gives two methods for sizing service-entrance conductors for residences or dwellings: the standard method and the alternate method.

The standard method requires that 3 watts per square foot be utilized for the general lighting load. The measurement is taken from the *outside* building dimensions, and each two-wire, small-appliance circuit is rated at 1500 watts. After these first loads are calculated, a demand factor must be applied; that is, the first 3000 watts is rated at 100 percent while the remaining watts of the general-lighting and small-appliance loads are rated at 35 percent.

The results of the calculation give the net computed load of the residence without electric range and major appliances. Therefore, all major appliances must be listed by their nameplate ratings, and the calculated load for the electric range, as described in Chapter 9, must be given. All figures are totalled to find the total calculated load in watts. This figure is divided by the phase-to-phase voltage to determine the load in amperes so that the service-entrance conductors may be sized according to Article 310 of the NEC.

Example 11–1

A single-level residence measures 24 feet by 55 feet and requires four small-appliance circuits; an 8000-watt electric-range circuit; a 4500-watt electric water heater circuit, a 4500-watt electric clothes-dryer circuit; and a 5600-watt air-conditioning circuit. Determine:

a   The total calculated load

b   The total load in amperes

c   The required wire size of THW copper

d   The required wire size of THW aluminum

Solution

a   The square footage must first be determined:

$$24 \times 55 = 1320 \text{ square feet}$$

Then establish the following table:

General lighting load:	
1320 square feet at 3 watts	3960 watts
Appliance circuits:	
4 at 1500	6000
Total	9960 watts
Application of demand factor [Table 220–4(b) of the NEC]	
3000 watts at 100 percent	3000 watts
9960-3000 watts = 6960 at 35 percent	2436

Net computed load without range and major appliances	5436 watts
Electric range (see NEC Table 220–5)	8000
Water heater	4500
Clothes dryer	4500
Air conditioner	5600
Total calculated load	28,036 watts

b $\text{Amperes} = \dfrac{28,036}{240 \text{ (volts)}} = 116.82$

c   From Tables in NEC Article 310, No. 2 AWG, THW copper wire will safely carry 115 amperes.

d   From Tables in NEC Article 310, 1/0 aluminum THW conductors will carry 120 amperes.

In the optional method of calculating residential service-entrance conductor sizes, the general lighting and small appliance loads are determined in the same manner as in the standard method. However, no diversity is taken until all the other loads have been determined. Furthermore, no diversity is allowed on the electric range: the nameplate rating must be figured in totalling the loads. All other major appliances are listed by their nameplate ratings and, all the loads are totalled. The load calculation by the optional method of the residence used to demonstrate the standard method follows:

General lighting load:	
1320 square feet at 3 watts	3960 watts
Small appliance load	
4 circuits at 1500	6000
Electric range (nameplate rating)	12,000
Water heater	4500
Clothes dryer	4500
Air conditioner	5600
Total connected load	36,560 watts

The first 10,000 watts (10 kW) must be rated at 100 percent (Article 220–7). The remaining load is calculated on the basis of a 40-percent diversity. Therefore,

First 10 kW at 100 percent	10,000 watts
Remaining load at 40 percent	
(36,560 − 10,000 = 26,560 × 0.40 =)	10,624
Total calculated load	20,624 watts

To find the total load in amperes, divide the total calculated wattage by the phase-to-phase voltage.

$$\frac{20,624}{240} = 85.9 \text{ amperes}$$

Table 310–19 of the NEC states that size No. 2 AWG aluminum conductors with THW insulation will safely carry the load. Footnote "c" of this table states that "for three-wire single-phase service the allowable ampacity of RH, RHH, RHW, and THW aluminum conductors is: No. 2—100 amp., No. 1—110 amps." This is good because no modern residence should be supplied with less than 100-ampere service—regardless of the calculated load.

Sizing the Neutral Conductor

The NEC states that the neutral conductor must be of sufficient size to carry the maximum *unbalanced* load determined by Section 220–4. The maximum unbalanced load is the maximum connected load between the neutral and any one ungrounded (phase) conductor. Therefore, no 240-volt, two-wire circuit will need a neutral conductor, and loads for these circuits do not have to be included in sizing the neutral conductor. This permits the neutral conductor to be somewhat smaller than the phase conductors.

The neutral conductor for the residence used in the previous service-entrance calculations should be sized by the following precedure.

General lighting and small-appliance load after applying demand factor (standard method)	9960 watts
Range load, 8000 watts at 70 percent (NEC Table 220–11)	5600
Total	15,560 watts

Then:

$$\frac{15,560}{240} = 64.83 \text{ amperes}$$

Based on the above calculation, the NEC permits the installation of a neutral conductor capable of carrying 64.83 amperes, so long as the wire size is not smaller than No. 4 copper or No. 6 aluminum. According to Table 310–79 of the NEC, this requires a No. 4 AWG conductor with THW insulation.

Sizing Services for Other than Residential Occupancies

Service-entrance conductors for all occupancies besides residences require the following procedure:

1   Determine the total load on all circuits feeding convenience outlets, based on 150 watts per receptacle (300 watts for each duplex receptacle).

2   Determine the total lighting load by listing the actual wattage of each fixture (where practical) or by Table 220–2(a) of the NEC.

3   List all pieces of electrical equipment along with their nameplate rating in watts.

4   Total all loads and divide by the phase-to-phase voltage to determine the load in amperes.

5   Select a wire size from the Table in Section 310 of the NEC.

Example 11–2

A small store building will have 31 duplex receptacles for general use; a 4500-watt, hot-water heater; a 47,000-watt, heat pump; and an undetermined amount of lighting. The building is 20 feet wide and 60 feet long. Find:

a   The calculated load in watts

b   The calculated load in amperes

c   The size of copper THW service-entrance conductors

d   The size of aluminum service-entrance conductors

Solution

a   Since the building is 20 feet wide and 60 feet long, it contains (20 × 60 =) 1200 square feet. From Table 220–2(a) of the NEC, the general lighting load for a store building should be based on 3 watts per square feet. Therefore, the estimated lighting load will be 1200 × 3 = 3600 watts. Thirty-one duplex receptacles at 300 watts = 9300 watts. The loads of the water heater and the heat pump total 51,500 watts. Therefore, the total calculated load for the building is:

General lighting	3600 watts
Convenience outlets	9300
Major appliances	51,500
	64,400 watts

b   The total load in amperes is $\dfrac{64,400}{240}$ 268 amperes.

c   From NEC Table, 300-MCM copper THW conductors safely carry 285 amperes.

d   From NEC Table 310–19, 400-MCM aluminum THW conductors safely carry 270 amperes.

There is no provision for a diversity factor for this type of building.

Sizing Grounding Conductors

Article 250 of the NEC covers general requirements for grounding and bonding electrical services. Table 11–1 gives the proper sizes of grounding conductors for various sizes of electric services.

Table 11–1   GROUNDING ELECTRODE CONDUCTOR FOR GROUNDED SYSTEMS

Size of largest service-entrance conductor or equivalent for parallel conductors		Size of grounding electrode conductor	
Copper	Aluminum or copper-clad aluminum	Copper	Aluminum or copper-clad aluminum
2 or smaller	0 or smaller	8	6
1 or 0	2/0 or 3/0	6	4
2/0 or 3/0	4/0 or 250 MCM	4	2
Over 3/0 thru 350 MCM	Over 250 MCM thru 500 MCM	2	0
Over 350 MCM thru 600 MCM	Over 500 MCM thru 900 MCM	0	3/0
Over 600 MCM thru 1100 MCM	Over 900 MCM thru 1750 MCM	2/0	4/0
Over 1100 MCM	Over 1750 MCM	3/0	250 MCM

11–2   SELECTING SERVICE EQUIPMENT

Service entrance equipment is usually grouped at one centralized location. Feeders run to various locations to feed heavy-loaded electrical equipment and subpanels, which are located in a building to keep the length of the branch-circuit raceways at a practical minimum for operating efficiency and in order to cut down on cost.

The main service-disconnecting means will sometimes be made up on the job by assembling individually enclosed fused switches on a length of metal auxiliary gutter, as shown in Figure 11–3. The various components are connected by means of short conduit nipples, in which the insulated conductors are installed. In other cases, the main disconnect and feeder overcurrent devices are enclosed in factory-assembled panelboard; the entire assembly is commonly called a main distribution panelboard.

In the selection of service-entrance equipment, the following factors must be taken into consideration.

Figure 11–3   INDIVIDUALLY ENCLOSED FUSE SWITCHES ON A LENGTH OF METAL AUXILIARY GUTTER.

1   Service characteristics available from the local utility company

2   Total connected load

3   Total demand of electrical load

4   Capacity for present and future loads

5   Requirements of the NEC, local ordinances, and the utility company feeding the building

6   Type and cost of equipment

7   Availability of equipment

Article 230–70 of the NEC requires that a service be provided with a disconnecting means for all conductors in the building or structure from the service-entrance conductors. This disconnecting means should be located at or near the point where the service-entrance conductors enter the building. The NEC also requires that all single feeders and branch circuits be provided with a means of individual disconnection from the source of supply. The disconnecting means must be located at a readily accessible point, either inside or outside the building, and adequate access and working space must be provided all around the disconnecting means.

Overcurrent protection is required both at the main source and for all individual feeders and branch circuits in order to protect the electrical installation against ground faults and overloads.

Fusible Service Equipment

Safety switches or fuse blocks in panelboards are rated at 15, 20, 60, 100, 200, 400, 600 etc. amperes. There are no in-between ratings, but fuses may be installed at any rating below the switch rating. For example, if a building requires a 350-ampere service, a 400-ampere switch will have to be installed. However, fuses rated at 350 amperes can be installed, allowing for the installation of conductors rated at 350 amperes.

Fusible panelboards, containing main fuse blocks and feeder and branch-circuit fuse spaces for 2 to 42 circuits, are available. A panelboard requiring more than 40 circuits on a single-phase service or 42 spaces on a three-phase service will have to be made into two or more panelboards.

Circuit Breakers

A circuit breaker resembles an ordinary toggle switch and is currently the most widely used means of overcurrent protection. On an overload, the circuit breaker opens itself or "trips." In a tripped position, the handle jumps to the middle position (Figure 11–4). To reset, turn the handle to the OFF position and push it as far as it will go beyond this position; then turn it to the ON position.

One, single-pole circuit breaker protects a 120-volt circuit or a 277-volt circuit on 480/277 Y systems. A double-pole circuit breaker is used to protect a 208- to 240-volt, two-wire circuit. A three-pole, circuit breaker protects three-phase circuits from 208 to 480 volts.

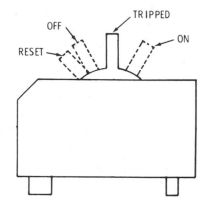

Figure 11– 4   POSITIONS OF A CIRCUIT BREAKER HANDLE IN "ON," "OFF," AND "TRIP" POSITIONS.

Circuit breaker enclosures are manufactured in several types. Some contain only branch-circuit breakers while others contain a main-circuit breaker in addition to branch-circuit breakers. Another type provides only a main-disconnecting means and protection for heavy feeder circuits. Enclosures that contain only a single-, double-, or three-pole circuit breaker are also available.

Circuit breakers are usually either the plug-in type or the bolt-in type. When the plug-in type is used, the cabinets are usually sold without the circuit breakers but contain an arrangement of bus bars so that the user can select whatever combination of circuit breakers is required and then plug them into the bus-bar arrangement. The bolt-in type is usually (although not always) factory assembled, and the designer or electrician must therefore know the loads of the various circuits so that the panels can be made up to suit the requirements.

If a circuit breaker panel contains six or less circuit breakers, it is permissible to eliminate a main-disconnecting means, provided the breakers are rated at more than 20 amperes.

Service equipment and its overcurrent protection devices must have short-circuit current rating equal to or greater than the available short-circuit current at its supply terminal (Article 230–98). Furthermore, ground-fault protection must be provided for grounded wye (three-phase) services or more than 150 volts to ground, but not exceeding 600 volts phase-to-phase for any service disconnecting means rated 1000 amperes or more. The ground-fault protection may consist of overcurrent devices or a combination of overcurrent devices and current transformers or other equivalent protection equipment, which operates to cause the service disconnecting means to open all ungrounded conductors of the faulted circuit at fault current values of 1200 amperes or more.

QUESTIONS

Answer the following questions by filling in the blanks.

1   All components between the point of termination of an overhead service drop and the building main-disconnecting device are known as the _____ _____.

2   A residential service entrance, in most cases, can be supplied through only _____ set of service-entrance conductors.

3   Service conductors must have adequate _____ to conduct the current for the loads supplied safely.

4   The neutral conductor must be of sufficient size to carry the maximum _____ load.

5   Article _____ of the NEC covers general requirements for grounding and bonding electrical services.

6   Service-entrance conductors of 500-MCM copper require a size _____ grounding conductor.

7   The NEC requires that a service, single feeders, and branch circuits be provided with a means of _____ from the source of supply

8   If a building requires a 175-ampere service, the fusible safety switch used has to be rated for _____ amperes.

9   Any panelboard requiring more than 40 circuits on a single-phase service has to be made into _____ or more panelboards.

10   To reset a circuit breaker, turn the handle to the _____ position and push it as far as it will go beyond this position. Then turn it to the _____ position.

12

SPECIAL INSTALLATIONS

12-1   HAZARDOUS LOCATIONS

Any area in which the atmosphere or a material in the area is such that the arcing of operating electrical contacts, components, and equipment may cause an explosion or fire is considered as a hazardous location. In all such cases, explosion-proof equipment, raceways, and fittings are used to provide an explosion-proof wiring system.

Hazardous locations have been classified in the NEC into certain class locations. Various atmospheric groups have been established on the basis of the explosive character of the atmosphere for the testing and approval of equipment for use in the various groups.

Class I locations   Those locations in which flammable gases or vapors may be present in the air in quantities sufficient to produce explosive or ignitible mixtures are classified as Class I locations. Examples of such locations are interiors of paint spray booths where volatile, flammable solvents are used, inadequately ventilated pump rooms where flammable gas is pumped, and drying rooms for the evaporation of flammable solvents.

Class II locations   Class II locations are those which are hazardous because of the presence of combustible dust. Class II, Division 1 locations are areas where combustible dust, under normal operating conditions, may be present in the air in quantities sufficient to produce explosive or ignitible mixtures; examples are working areas of grain handling and storage plants and rooms containing grinders or pulverizers. Class II, Division 2 locations are areas where dangerous concentrations of suspended dust are not likely, but where dust accumulations *might* form.

Class III locations   These locations are those areas which are hazardous because of the presence of easily ignitible fibers or flyings, but such fibers and flyings are not likely to be in suspension in the air in these locations in quantities sufficient to produce ignitible mixtures. Such locations usually include some parts of rayon, cotton, and textile mills; clothing manufacturing plants; and woodworking plants.

The wide assortment of explosion-proof equipment now available makes it possible to provide adequate electrical installations under any of these hazardous conditions. However, the electrician must be thoroughly familiar with all NEC requirements, and know what fittings are available, how to install them properly, and where and when to use the various fittings.

The usual working drawings for a hazardous area are drawn the same as the layout of an electrical system for a nonhazardous area—the only distinction is a note on the drawings stating that the wiring in this particular area shall conform to the NEC requirements for hazardous locations. The designer will sometimes add the letters ''EXP'' next to all the symbols of the outlet that are to be explosion proof. However few engineers or draftsmen detail their drawings for hazardous areas sufficiently for the electricians to proceed with the installation without additional study and layout work on the job site. Therefore, an electrician *must* be familiar with the layout and installation procedures before attempting such an installation. For other than very simple installations, it may be advisable to make rough, detailed wiring layouts of the proposed installation, even if they are merely sketches on the original working drawings.

A partial floor plan for a hazardous area is shown in Figure 12–1. Note that few installation details are given. The drawing in Figure 12–2 is more detailed and gives adequate information for experienced electricians to complete the installation without additional questions or help.

When an electrical system in a hazardous location is designed or installed, the type of building structure and finish must be considered. If the building is under construction, this information may be obtained from the architectural drawings and specifications. If the installation is made in an existing building, a preliminary job-site investigation is often necessary. The location of the explosion-proof outlets, whether concealed or exposed, and the class of hazardous locations should appear in the electrical drawings and specifications. If such information is not provided, the contractor or electrician will have to determine this information for himself—from the architect, owner, local inspection authority, etc.

In general, rigid metallic conduit is required for all hazardous locations,

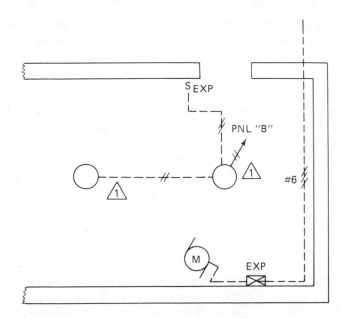

Figure 12–1   PARTIAL FLOOR PLAN OF AN AREA CLASSIFIED AS HAZARD-OUS.

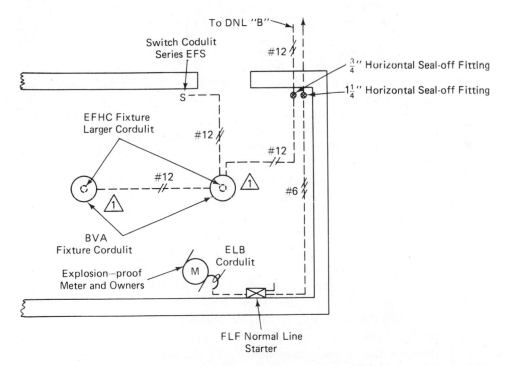

Figure 12–2   A MORE DETAILED DRAWING OF THE AREA SHOWN IN FIG-URE 12–1.

except for special flexible terminations and as otherwise permitted in the NEC. The conduit should be threaded with a standard conduit cutting die that provides 3/4-inch taper per foot. The conduit should be made up wrench tight in order to minimize sparking in the event fault current flows through the conduit system (NEC Article 500–1). Where it is impractical to make a threaded joint tight, a bonding jumper should be used. All boxes, fittings, and joints shall be threaded for connection to the conduit system and shall be an approved, explosion-proof type. Threaded joints shall be made up with at least five threads fully engaged. Where it becomes necessary to employ flexible connectors at motor or fixture terminals, flexible fittings approved for the particular Class location shall be used.

Seal-off fittings are required in conduit systems to prevent the passage of gases, vapors, or flames from one portion of the electrical installation to another through the conduit. For Class I, Division I locations, the NEC (Article 501–5) states that

> In each conduit run entering an enclosure for switches, circuit breakers, fuses, relays, resistors or other apparatus which may produce arcs, sparks or high temperatures, seals shall be placed as close as practicable and in no case more than 18 inches from such enclosures. There shall be no junction box or similar enclosure in the conduit run between the sealing fitting and the apparatus enclosure . . .
>
> In each conduit run of 2-inch size or larger entering the enclosure or fitting housing terminals, splices or taps, and within 18 inches of such enclosure or fitting . . .
>
> In each conduit run leaving the Class I, Division I hazardous area. The sealing fitting may be located on either side of the boundary of such hazardous area, but shall be so designed and installed that any gases or vapors which may enter the conduit system, within the Division 1 hazardous area, will not enter or be communicated to the conduit beyond the seal. There shall be no union, coupling, box or fitting in the conduit between the sealing fitting and the point at which the conduit leaves the Division 1 hazardous area . . .

Sealing compound shall be approved for the purpose, shall not be affected by the surrounding atmosphere or liquids, and shall not have a melting point of less than 200°F (93°C). Most sealing-compound kits contain a powder in a polyethylene bag within an outer container. To mix, remove the bag of powder, fill the outside container with water up to the marked line on the container, and pour in the powder and mix.

To pack the seal off, remove the threaded plug or plugs from the fitting and insert the asbestos fiber supplied with the packing kit. Tamp the fiber between the wires and the hub before pouring the sealing compound into the fitting. Then pour in the sealing cement and reset the threaded plug tightly. The fiber packing prevents the sealing compound (in the liquid state) from entering the conduit lines.

The seal-off fittings in Figure 12–3 are typical of those used. The type in Figure 12–3a is for vertical mounting and is provided with a threaded, plugged

Figure 12–3a   TYPICAL SEAL-OFF FITTING FOR VERTICAL MOUNTING.

Figure 12–3b   A SEAL-OFF FITTING WITH AN ADDITIONAL PLUG OPENING IN THE LOWER HUB TO FACILITATE PACKING FIBER AROUND THE CONDUCTORS TO FORM A DAM FOR THE SEALING CEMENT.

opening into which the sealing cement is poured. The seal-off in Figure 12–3b has an additional plugged opening in the lower hub to facilitate packing fiber around the conductors in order to form a dam for the sealing cement.

Most other explosion-proof fittings are provided with threaded hubs for securing the conduit as described previously. Typical fittings include switch and junction boxes, conduit bodies, union and connectors, flexible couplings, explosion-proof lighting fixtures, receptacles, and panelboard and motor starter enclosures. A practical representation of these and other fittings is shown in Figure 12–4.

12–2   GARAGES AND SIMILAR LOCATIONS

Garages and similar locations where volatile or flammable liquids are handled or used as fuel in self-propelled vehicles (including automobiles, buses, trucks, and tractors) are not usually considered critically hazardous locations. However, the entire area up to a level 18 inches above the floor is considered a Class I, Division 2 location, and certain precautionary measures are required by the NEC. Likewise, any pit or depression below floor level shall be considered a Class I, Division 2 location, and the pit or depression may be judged as Class I, Division 1 location if it is unvented.

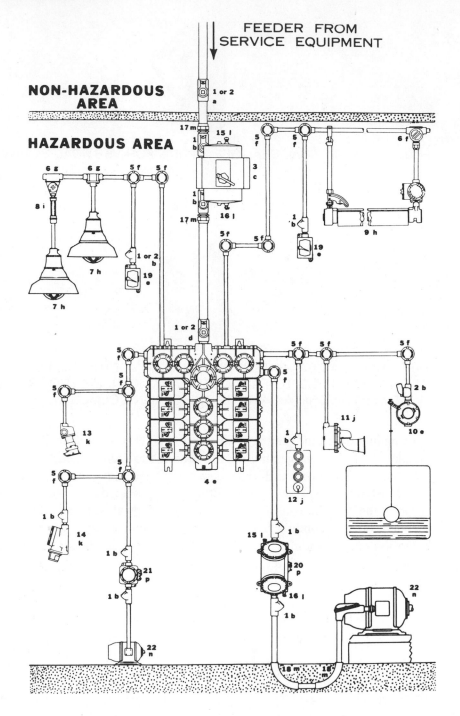

Figure 12–4 A PRACTICAL REPRESENTATION OF SEVERAL EXPLOSION-PROOF FITTINGS.

Normal raceway (conduit) and wiring may be used for the wiring method above this hazardous level, except where conditions indicate that the area concerned is more hazardous than usual. In this case, the applicable type of explosion-proof wiring may be required.

Approved seal-off fittings should be used on all conduit passing from hazardous area to non-hazardous area. The requirements set forth in NEC Sections 501–5 and 501–5(b)(2) shall apply to horizontal as well as vertical boundaries of the defined hazardous areas. Raceways embedded in a masonry floor or buried beneath a floor are considered to be within the hazardous area above the floor *if any connections or extensions* lead into or through such an area.

12–3   THEATERS

The NEC recognizes that hazards to life and property due to fire and panic exist in theaters, cinemas, etc. The NEC therefore requires certain precautions in these areas in addition to those of commercial installations. These requirements include:

1   Proper wiring of motion picture projection rooms (Article 540)

2   Heat-resistant, insulated conductors for certain lighting equipment (Article 520–43(b))

3   Adequate guarding and protection of the stage switchboard and proper control and overcurrent protection of circuits (Article 520–22)

4   Proper type and wiring of lighting dimmers (Articles 520–53(e) and 520–25)

5   Use of proper types of receptacles and flexible cables for stage lighting equipment (Article 520–45)

6   Proper, stage flue, damper control (Article 520–49)

7   Proper dressing-room wiring and control (Articles 520–71, 72, and 73)

8   Fireproof projection rooms with automatic projector port closures, ventilating equipment, emergency lighting, guarded work lights, and proper location of related equipment (Article 540)

Outdoor or drive-in motion picture theaters do not present the inherent hazards of enclosed auditoriums. However, the projection rooms must be properly ventilated and wired for the protection of the operating personnel.

12–4   HOSPITALS

Hospitals and other health care facilities fall under Article 517 of the NEC. To better understand the provisions set forth in this article, the reader should understand the following definitions:

Alternate power source—One or more generator sets (See Section 15–1) intended to provide power during the interruption of the normal electrical service or the public utility electrical service intended to provide power during interruption of service, normally provided by generating facilities on the premises.

Anesthetizing location—Any area intended for the administration of any flammable or nonflammable inhalation anesthetic agents in the course of examination or treatment, including operating rooms, delivery rooms, emergency rooms, anesthetizing rooms, corridors, and utility rooms.

Anesthetizing-location receptacle—A receptacle designed to accept the attachment plugs recognized for use in such locations.

Continuous power system—An electrical system independent of the alternate source that supplies power without appreciable interruption (one cycle or less).

Critical branch—A subsystem of the emergency system that consists of feeders and branch circuits supplying energy to task illumination and selected receptacles serving areas and functions related to patient care and that can be connected to alternate power sources by one or more transfer switches.

Critical patient care area—A section (rooms, wards, or portions of wards) designated for the treatment of critically ill patients.

Electrically susceptible patient—A patient being treated with an externalized electric conductor, such as a probe, catheter, or other electrode connected to the heart.

Electrically susceptible patient area—A location in a health care facility where electrically susceptible patients are cared for collectively.

Emergency system—A system of feeders and branch circuits meeting the requirements of Article 700, connected to alternate power sources by a transfer switch and supplying energy to an extremely limited number of prescribed functions vital to the protection of life and patient safety, with automatic restoration of electrical power within 10 seconds of power interruption.

Equipment system—A system of feeders and branch circuits arranged for delayed, automatic, or manual connection to the alternate power source that serves primarily three-phase power equipment.

Essential electrical systems—Systems comprised of alternate sources of power, transfer switches, overcurrent protective devices, distribution cabinets, feeders, branch circuits, motor controls, and all connected electrical equipment designed to provide designated areas with continuity of electrical service during disruption of normal power sources and to minimize the interruptive effects of disruption within the internal wiring system.

Flammable anesthetics—Gases or vapors such as fluroxene, cyclopropane, divinyl ether, ethyl chloride, ethyl ether, and ethylene, which may form flammable or explosive mixtures with air, oxygen, or reducing gases such as nitrous oxide.

Flammable anesthetizing location—Any operating room, delivery room, anesthetizing room, corridor, utility room, or any other area used or intended for the application of flammable anesthetics.

Health care facilities—Buildings or parts of buildings that contain, but are not limited to, hospitals, nursing homes, extended-care facilities, clinics, and medical and dental offices, whether fixed or mobile.

Immediate restoration of service—Automatic restoration of operation with an interruption of not more than 10 seconds as applied to those areas and functions served by the emergency system, except for areas and functions for which Article 700 makes specific provisions.

Intensive care units—Groups of beds, rooms, or wards specifically designated to provide intensive care for critically ill patients and intended to be staffed and organized for such service, distinct from surgical or obstetrical recovery units forming a part of a surgical or obstetrical suite.

Life safety branch—A subsystem of the emergency system consisting of feeders and branch circuits, meeting the requirements of Article 700 and intended to provide adequate power needs to insure safety to patients and personnel and that can be connected to alternate power sources by one or more transfer switches.

Life support branch—The life support branch of the emergency system supplies power centers in electrically susceptible patient locations.

Line isolation monitor—A test instrument designed to check continually the balanced and unbalanced impedance from each line of an isolated circuit to ground and equipped with a built-in test circuit to exercise the alarm without adding to the leakage current hazard.

Nurses' stations—Areas intended to provide a center of nursing activity for a group of nurses working under one nurse supervisor and serving bed patients, where the patient calls are received, nurses are dispatched, nurses' notes are written, inpatient charts are prepared, and medications are prepared for distribution to patients. Where such activities are carried on in more than one location within a nursing unit, all the separate areas are considered a part of the nurse's station.

Probable failure—One or more failures of the following:

1 Any single component

2 Any components that might fail without detection during normal use, including interruption of the grounding conductor

3 Any components that might fail as a result of the failure of any or all of the above components

Reference grounding bus, patient—The terminal grounding bus that serves as the single focus for grounding the electrical equipment connected to an individual patient or for grounding the metal or conductive furniture or other equipment within reach of the patient or a person who may be touching the patient.

Reference grounding bus, room—The terminal grounding bus that serves as the single focus for grounding the patient reference grounding buses and all other metal or conductive furniture, equipment, or structural surfaces in the room.

Task illumination—Provision for the minimum lighting required to carry out necessary tasks in the described areas, including safe access to supplies, equipment, and exits.

Part B of Article 517 (NEC) covers the general wiring systems of health care facilities. Part C covers essential electrical systems for hospitals. Part D gives the performance criteria and wiring methods to minimize shock hazards to patients in electrically susceptible patient areas. Part E covers the requirements for electrical wiring and equipment used in inhalation anesthetizing locations.

With the widespread use of X-ray equipment of varying types in health care facilities, electricians are often required to wire and connect equipment such as discussed in Article 660 of the NEC. Conventional wiring methods are used, but provisions should be made for 50- and 60-ampere receptacles for medical X-ray equipment (Section 660–3b).

12–5 AIRPORT HANGARS

Buildings used for storing or servicing aircraft in which gasoline, jet fuels, or other volatile flammable liquids or gases are used fall under Article 513 of the NEC. In general, any pit or depression below the level of the hangar floor is considered to be a Class I, Division 1 location. The entire area of the hangar including any adjacent and communicating area not suitably cut off from the hangar is considered to be a Class I, Division 2 location up to a level of 18 inches above the floor. The area within five feet horizontally from aircraft power plants, fuel tanks, or structures containing fuel is considered to be a Class I, Division 2 hazardous location; this area extends upward from the floor to a level five feet above the upper surface of wings and of engine enclosures.

Adjacent areas in which hazardous vapors are not likely to be released, such as stock rooms and electrical control rooms, should not be classed as hazardous when they are adequately ventilated and effectively cut off from the hangar itself by walls or partitions. All fixed wiring in a hangar not within a hazardous area are defined in Section 513–2 must be installed in metallic raceways or shall be Type MI or Type ALS cable; the only exception is wiring in nonhazardous locations as defined in Section 513–2 (d) which may be of any type recognized in Chapter 3 of this code.

QUESTIONS

Read each question carefully before answering the questions. A copy of the latest NEC will be helpful.

1   How should a motor be protected if it must operate in a paint room that contains explosive vapors?

2   What part of outdoor or drive-in theaters would fall under Chapter 5 of the NEC?

3   _____ insulated conductors are required by NEC Article 520–43 (b) for certain lighting installations in theaters.

4   Part _____ of Article 517 covers the general wiring systems of health care facilities.

5   What step is required before pouring the sealing cement into the threaded hub of an explosion-proof, seal-off fitting?

6   What rigid galvanized conduit in an explosion-proof wiring system is used, what is the minimum number of threads that should be taken up at all terminals, fittings, etc?

7   The main purpose of an explosion-proof wiring system is to maintain all electrical wiring and equipment in _____ enclosures and to insure that the conduit connecting the equipment will be blocked off by _____ that prevent the passage of gas, and therefore explosions, through the raceway system.

8   In locations where hazardous dusts are present, all electrical equipment should be _____ so there will be no explosions of dust within the apparatus or its control devices.

9   The area _____ inches above a garage floor is considered to be a Class I, Division 2 location.

10   Define a Class III, hazardous location as specified in the NEC.

13

MOBILE HOME AND TRAILER PARK INSTALLATIONS

In recent years, mobile home and trailer parks have increased in number to the point where the NFPA has found it necessary to add additional requirements to the NEC. In the 1971 edition, Article 550, "Mobile Homes and Mobile Home Parks," and Article 551, "Recreation Vehicles and Parks," were added. Subsequent editions have retained and supplemented these articles.

The pertinent provisions of these articles cover the electrical conductors and the equipment installed within or on mobile homes and recreation vehicles, as well as the means of connecting the units to a supply of electricity.

13-1 DEFINITIONS

The following definitions should be learned to understand fully the material presented here which should be understood before attempting an installation on any facility for mobile homes, travel trailers, or any type of recreational vehicle.

Air-conditioning or comfort-cooling equipment—All that equipment intended or installed for the purpose of processing the treatment of air in order to control simultaneously its temperature, humidity, cleanliness, and distribution to meet the requirements of the conditioned space.

Camping trailer—A vehicular, portable unit mounted on wheels and constructed with collapsible, partial side walls that fold for towing by another vehicle and unfold at the campsite to provide temporary living quarters for recreational, camping, or travel use.

Converter—A device that changes electrical energy from one form to another, for example, from alternating current to direct current.

Dead front—As applied to switches, circuit breakers, switchboards, and distribution panelboards, a front so designed, constructed, and installed that no current-carrying parts are exposed.

Disconnecting means—The necessary equipment, usually consisting of a circuit breaker or switch and fuses and their accessories, located near the point of entrance of supply conductors in a recreational vehicle and intended to constitute the means cutting off the supply to that recreational vehicle. Receptacles used as disconnecting means must be accessible (as applied to wiring methods) and capable of interrupting their rated current without hazard to the operator.

Distribution panelboard—A single panel or group of panel units designed for assembly in the form of a single panel; including buses, and with or without switches and/or automatic overcurrent protection devices for the control of light, heat, or power circuits of small, individual as well as aggregate capacity; designed to be placed in a cabinet or cutout box placed in or against a wall or partition and accessible only from the front.

Feeder assembly—The overhead or under-chassis feeder conductors, including the grounding conductor, together with the necessary fittings and equipment or a power-supply cord approved for mobile home use, designed for the purpose of delivering energy from the source of electrical supply to the distribution panelboard within the mobile home.

Mobile home—A factory-assembled structure or structures equipped with the necessary service connections and made so as to be readily movable as a unit or units on their own running gear and designed to be used as dwelling unit(s) without a permanent foundation.

Mobile home accessory building or structure—Any awning, cabana, ramada, storage cabinet, carport, fence, windbreak, or porch established for use of the occupant of the mobile home on a mobile home lot.

Mobile home lot—A designated portion of a mobile home park designed for the accommodation of one mobile home and its accessory buildings or structures for the exclusive use of its occupants.

Mobile home park—A contiguous parcel of land that is used for the accommodation of occupied mobile homes.

Mobile home service equipment—The equipment containing the disconnecting means, overcurrent protection devices, and receptacles or other means for connecting a mobile home feeder assembly.

Motor home—A vehicular unit built on a self-propelled motor vehicle chassis, primarily designed to provide temporary living quarters for recreational, camping, or travel use.

Park electrical wiring system—All of the electrical wiring, fixtures, equipment, and appurtenances related to electrical installations within a mobile home park, including the mobile home service equipment.

Power-supply assembly—The conductors (including the grounding conductors) insulated from one another, the connectors, attachment plug caps, and all other fittings, grommets, or devices installed for the purpose of delivering energy from the source of electrical supply to the distribution panel within the recreational vehicle.

Recreational vehicle—A vehicular-type unit primarily designed as temporary living quarters for recreational, camping, or travel use that either has its own motive power or is mounted on or is drawn by another vehicle. The basic entities are travel trailer, camping trailer, truck camper, and motor home.

Transformer—A device that raises or lowers the voltage of alternating current of the original source.

Travel trailer—A vehicular portable unit mounted on wheels of such size or weight as not to require special highway movement permits when drawn by a motorized vehicle, of body width of no more than 8 feet and length of no more than 32 feet when factory equipped for the road; primarily designed and constructed to provide temporary living quarters for recreational, camping, or travel use.

Truck camper—A portable unit, designed to be loaded onto, or affixed to, the bed or chassis of a truck; constructed to provide temporary living quarters for recreational, camping, or travel use. Truck campers are of two basic types:

a Slide-in camper   A portable unit designed to be loaded onto and unloaded from the bed of a pickup truck; constructed to provide temporary living quarters for recreational, travel, or camping use.

b Chassis-mount camper   A portable unit designed to be affixed to a truck chassis; constructed to provide temporary living quarters for recreational, travel, or camping use.

13–2 SIZING ELECTRICAL SERVICES FOR MOBILE HOMES

The electrical service for a mobile home may be installed either underground or overhead, but the point of attachment must be a pole or power pedestal located adjacent to but *not* mounted on or in the mobile home. The power supply to the mobile home itself is provided by a feeder assembly consisting of not more than three mobile home power cords—each rated for at least 50 amperes or a permanently installed circuit.

The NEC gives specific instructions for determining the size of the supply-cord and distribution-panel load for each feeder assembly for each mobile home. The calculations are based on the size of the mobile home, the small-appliance circuits, and other electrical equipment that will be connected to the service.

Lighting loads are computed on the basis of the mobile home's area: width times length (outside dimensions exclusive of coupler) times 3 watts per square foot.

$$\text{Length} \times \text{width} \times 3 = \underline{\hspace{2cm}} \text{ lighting watts}$$

Small appliance loads are computed on the basis of the number of circuits times 1500 watts for each 20-ampere appliance, receptacle circuit.

Number of circuits $\times$ 1,500 = _____ small appliance watts

The sum of the two loads gives the total load in watts. However, there is a diversity (demand) factor that may be applied to this total in sizing the service and power cord. The first 3000 watts (obtained from the previous calculation) is rated at 100 percent. The remaining watts should be multiplied by a demand factor of 0.35 (35 percent). The total wattage so obtained is divided by the feeder voltage to obtain the service size in amperes.

If other electrical loads are to be used in the mobile home, the nameplate rating of each must be determined and entered in the summation. Therefore, to determine the total load for a mobile home power supply, calculate:

1   Lighting and small appliance load, as discussed previously.

2   Nameplate amperes for motors and heater loads, including exhaust fans, air conditioners, and electric heaters. Since air conditioners and heaters will not operate simultaneously, only the larger of the two needs to be included in the total load figures. Multiply the largest motor nameplate rating by 1.25 and add the answer in the calculations.

3   Total of nameplate amperes for any garbage disposals, dishwashers, electric water heaters, clothes dryers, cooking units, etc. Where there are more than three of these appliances, use 75 percent of the total load.

4   The amperes for free-standing ranges (as distinguished from separate ovens and cooking units) by dividing values shown in Table 13–1 by the voltage between phases.

5   The anticipated load if outlets or circuits are provided for other than factory-installed appliances.

Table 13–1   POWER DEMAND FACTORS FOR FREE-STANDING ELECTRIC RANGES

Nameplate rating	Use
10,000 watts or less	80 percent of rating
10,001 - 12,500 watts	8000 watts
12,501 - 13,500 watts	8400 watts
13,501 - 14,500 watts	8800 watts
14,501 - 15,500 watts	9200 watts
15,501 - 16,500 watts	9600 watts
16,501 - 17,500 watts	10,000 watts

To illustrate this procedure for determining the size of the electrical service and power cord for a mobile home, assume that a mobile home is 70 feet × 10 feet and has three, portable, appliance circuits; a 1200-watt air conditioner; a 200-watt, 120-volt exhaust fan; a 1500-watt water heater; and a 8000-watt electric range. The load is calculated as follows:

Lighting and small appliance load:	
Lighting load:	
70 feet × 10 feet × 3 watts/ft²	2100 watts
Small appliance load:	
1500 watts/circuit × 3 circuits	4500
	6600 watts
1st 3000 at 100 percent	3000 watts
Remainder	
6600 − 3000 = 3600 at 35 percent	1260
	4260 watts

$$\frac{4260 \text{ watts}}{240 \text{ volts}} = 17.75 \text{ amperes per phase}$$

Large appliance load:

1500 watts (water heater) ÷ 240 volts	6.25 A/line
200 watts (fan) ÷ 120 volts (one phase)	1.66 A/phase
1200 watts (air conditioner) ÷ 240 volts	5.00 A/line
8000 watts (range) × 0.8 (diversity factor) ÷ 240 volts	26.66 A/line

Summary:	Amperes per phase	
	A	B
Lighting and appliance	17.75	17.75
Water heater	6.25	6.25
Fan	1.66	—
Air conditioner	5.00	5.00
Range	26.66	26.66
Total	57.32	55.66

Based on the higher current for either phase, a 60-ampere power cord should be used to furnish electric power for the mobile home. The service should be rated for a minimum of 60 amperes and be fused accordingly.

13-3 TYPES OF EQUIPMENT

Weather-proof electrical equipment for mobile homes, mobile home parks, and similar outdoor applications are available from many sources. One of the major suppliers of weather-proof power outlets is Midwest Electric Products,

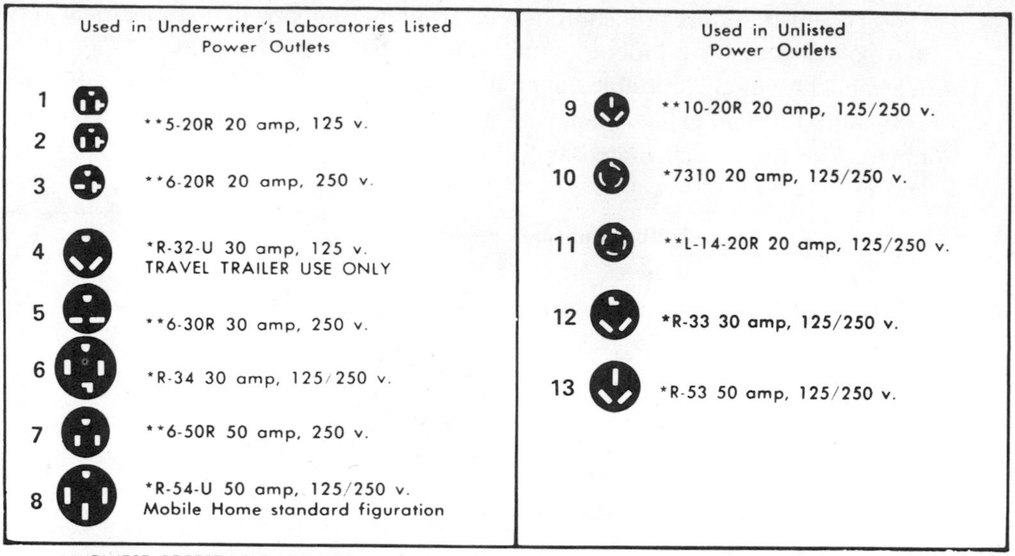

Used in Underwriter's Laboratories Listed Power Outlets	Used in Unlisted Power Outlets
1 ⬛ **5-20R 20 amp, 125 v.	9 ⬛ **10-20R 20 amp, 125/250 v.
2 ⬛	10 ⬛ *7310 20 amp, 125/250 v.
3 ⬛ **6-20R 20 amp, 250 v.	11 ⬛ **L-14-20R 20 amp, 125/250 v.
4 ⬛ *R-32-U 30 amp, 125 v. TRAVEL TRAILER USE ONLY	12 ⬛ *R-33 30 amp, 125/250 v.
5 ⬛ **6-30R 30 amp, 250 v.	13 ⬛ *R-53 50 amp, 125/250 v.
6 ⬛ *R-34 30 amp, 125/250 v.	
7 ⬛ **6-50R 50 amp, 250 v.	
8 ⬛ *R-54-U 50 amp, 125/250 v. Mobile Home standard figuration	

*MIDWEST RECEPTACLE NUMBER
**NEMA RECEPTACLE CONFIGURATION NUMBER

Figure 13–1   RECEPTACLE CONFIGURATIONS.

Inc., Mankato, Minnesota. The reader should obtain one of their free catalogs and study the many types of mobile home utility power outlets and service equipment manufactured by this company.

Receptacle configurations used in mobile home and recreation vehicle applications are shown in Figure 13–1. Receptacles 1, 2, 4, and 8 are the most commonly used in mobile home and recreation vehicle parks.

Power units for use in mobile homes and recreation vehicles vary in design with fuse or circuit-breaker projection, attached meter sockets, mounting and junction posts, special corrosion-resistant finish for ocean-side areas. The latter, of course, are for boats where the outlets are located along the docks and power is leased by the dock owners on a daily basis. In addition to the standard units, manufacturers build equipment to meet special requirements. Merely write the manufacturer, and they will supply you with a price and delivery date on the equipment.

Where more than one mobile home is to be fed, a power outlet and service equipment mounting cubicle (section shown in Figure 13–2) is ideal. The bus bars in this unit accommodate wire sizes to 600 MCM for a 400-ampere capacity. Therefore, either two 200-ampere units or four 100-ampere units may be mounted on the cubicle for feeding mobile home units underground. The main service should also enter from underground.

Standard 15-ampere or 20-ampere, 120-volt receptacles may be converted for use with standard 30-ampere, 120-volt travel trailer caps by use of the adapter shown in Figure 13–3.

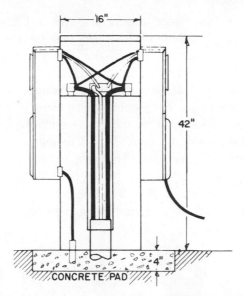

Figure 13–2 MOUNTING CUBICLE FOR ACCOMMODATING UP TO FOUR MOBILE HOMES.

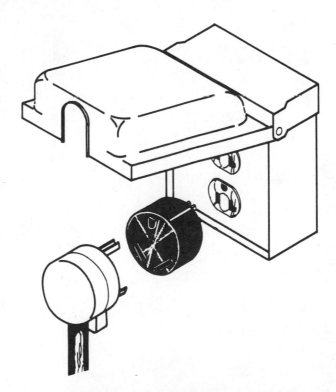

Figure 13–3 TRAVEL TRAILER ADAPTERS ARE AVAILABLE TO ADAPT THE STANDARD, 30-AMPERE, 125-VOLT, TRAVEL TRAILER CAP TO ANY STANDARD, 15-AMPERE OR 20-AMPERE-, 125-VOLT OUTLET.

The power outlet mounting post shown in Figure 13–4 is very popular for travel trailer parks and marinas. A cast-aluminum mounting base is provided to mount the power outlet on. The installer provides a length of 2 inch rigid conduit of the desired length. This is for underground installations. Note the conductors feeding in and out of the bottom of the conduit.

A side view of another mounting post and wire trough for travel trailer parks where the installation requires the power outlet to be mounted remote from the meter socket is shown in Figure 13–5. A mounting post for mobile home applications is shown in Figure 13–6. This post accommodates one or two combination meter/power units, back-to-back on the posts. When the power units are ordered from the factory prewired, the entire unit is ready to be secured in the ground and only the connections to the terminal bar have to be made.

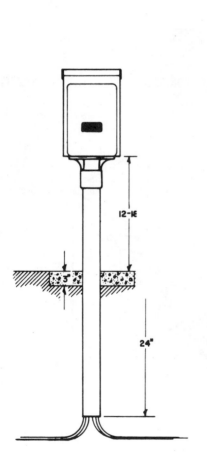

Figure 13–4 A POPULAR, POWER-OUTLET MOUNTING POST.

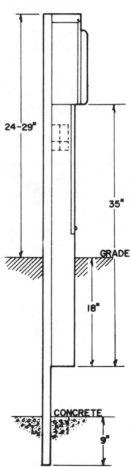

Figure 13–5 SIDE VIEW OF ANOTHER TYPE OF MOUNTING POST FOR TRAVEL TRAILER PARKS.

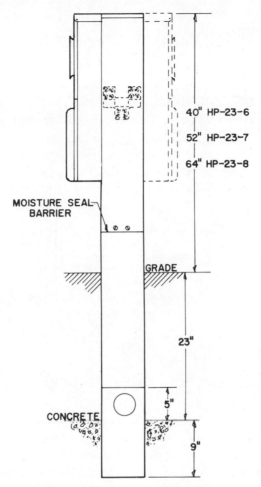

Figure 13–6   MOUNTING POST FOR MOBILE HOME APPLICATIONS.

13–4   INSTALLATION METHODS

There are certain requirements in the NEC as well as accepted methods for installing mobile home services and feeders. The electrician should become familiar with these prior to beginning an installation. Some of the more commonly used methods are discussed below.

A typical mobile home park service pole/meter installation is shown in Figure 13–7. This particular arrangement is designed to serve four mobile home units.

Before beginning an installation, the customer or the electrician performing the work should consult with the local power company for the method of serving the mobile home park and for the location of service-entrance poles. Power

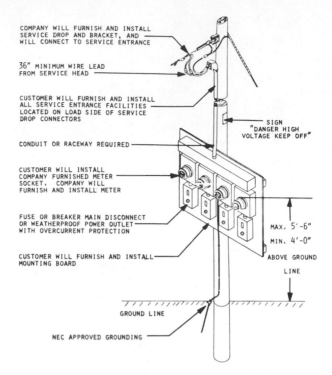

COMPANY WILL FURNISH AND INSTALL
SERVICE DROP AND BRACKET, AND
WILL CONNECT TO SERVICE ENTRANCE

36" MINIMUM WIRE LEAD
FROM SERVICE HEAD

CUSTOMER WILL FURNISH AND INSTALL
ALL SERVICE ENTRANCE FACILITIES
LOCATED ON LOAD SIDE OF SERVICE
DROP CONNECTORS

CONDUIT OR RACEWAY REQUIRED

CUSTOMER WILL INSTALL
COMPANY FURNISHED METER
SOCKET. COMPANY WILL
FURNISH AND INSTALL METER

FUSE OR BREAKER MAIN DISCONNECT
OR WEATHERPROOF POWER OUTLET
WITH OVERCURRENT PROTECTION

CUSTOMER WILL FURNISH AND INSTALL
MOUNTING BOARD

SIGN
"DANGER HIGH
VOLTAGE KEEP OFF"

MAX. 5'-6"

MIN. 4'-0"

ABOVE GROUND
LINE

GROUND LINE

NEC APPROVED GROUNDING

Figure 13–7 TYPICAL MOBILE HOME PARK SERVICE.

company regulations vary from area to area, but the power company will furnish and set the pole in most cases. However, the electrician must obtain permission from the power company before perfoming any work on facilities on the poles.

Once a definite plan has been settled upon, the electrician should put a piece of 1/2-inch thick plywood of sufficient size to hold the service equipment, i.e., a wire trough, meter bases, and weather-proof power outlets as shown in Figure 13–7. This piece of plywood should be primed with paint, and a final coat of wood preservant then applied. Two pieces of 2 inch × 4 inch timbers are spiked or otherwise secured to the sheet of plywood for reinforcement before the entire assembly is spiked to the pole. The wood backing should be arranged so that the meters will be no more than 5 feet 6 inches above the ground nor less than 4 feet when they are installed.

Up to six power outlets may be fed from one service without needing a disconnect switch to shut down the entire service. However, if more than four power outlets are assembled on one piece of plywood, the arrangement shown in Figure 13–7 will not provide adequate support. Another short pole should be installed at a distance from the service pole so that the sheet of plywood can be secured to both poles (Figure 13–8) for added support.

With the plywood backing secured in place, a wire trough, sized according to Article 374–5 of the NEC, should be installed at the very top of the board as shown in Figure 13–7. The wire trough (auxiliary gutter) should not contain more

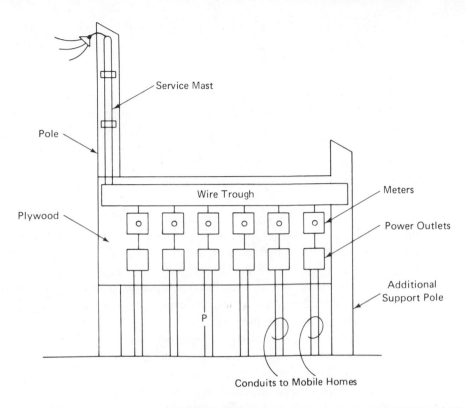

Figure 13–8 METHOD OF INSTALLING AN ADDITIONAL SHORT POLE AD-JACENT TO THE SERVICE POLE FOR ADDED SUPPORT.

than 30 current-carrying conductors nor should the sum of the cross-sectional areas of all contained conductors at any cross section exceed 20 percent of the interior cross-sectional area of the gutter. The auxiliary gutter should be approved for outdoor use.

The meter bases may usually be obtained from the power company but must be installed by the electrician. Once the entire installation is complete and has been inspected, the power company will install the meters. The connections of the meter bases to the wire trough are made with short, rigid conduit nipples using locknuts and bushings. Although straight nipples are often used for these connections, an offset nipple (Figure 13–9) usually does a better job.

Figure 13–9 OFFSET NIPPLE USED TO CONNECT PIECES OF ELECTRICAL EQUIPMENT.

Weather-proof fuse or circuit-breaker disconnects are installed directly under the meter bases—again by means of conduit nipples. A weather-proof, 50-ampere, mobile home power outlet with overcurrent protection is also used quite often.

The service mast comes next and should consist either of rigid metallic conduit or of EMT with weather-proof fittings. Once installed and secured to the pole with pipe straps, the service-entrance conductors may be pulled into the conduit and out into the wire trough. An approved weatherhead is then installed on top of the mast, and at least three feet of service conductors should be left for the power company to make their connections.

With the service-entrance conductors in place, meter taps are made to the service conductors in the trough. All such splices and taps made and insulated by approved methods may be located within the gutter when the taps are accessible by means of removable covers or doors on the wire trough. The conductors, including splices and taps, must not fill more than 75 percent of the gutter area (Article 374–8a). These taps must leave the gutter opposite their terminal connections, and conductors must not be brought in contact with uninsulated current-carrying parts of opposite polarity.

The taps in the auxiliary gutter go directly to the line side of the meter bases. Once secured, the load side of the meter bases are connected to the disconnects or power outlets. All wiring should be sized according to the NEC.

Most water supplies for mobile homes consist of PVC (plastic) water pipe and, therefore, cannot provide an adequate ground for the service equipment. In cases like these, a *grounding electrode,* such as a 3/4 inch × 8 foot ground round driven in the ground near the service equipment is used. A piece of bare copper ground wire is connected to the ground rod on one end with an approved ground clamp, and the other end is connected to the neutral wire in the auxiliary gutter. This wire must be sized according to Table 250–94a of the NEC.

When all the work is complete, the service installation should be inspected by the local electrical inspector. The power company should then be notified to provide final connection of their lines.

13–5 SIZING ELECTRICAL SERVICES AND FEEDERS FOR PARKS

A minimum of 75 percent of all recreation vehicle park lots with electrical service equipment must be equipped with both a 20-ampere, 125-volt receptacle and a 30-ampere, 125-volt receptacle. The remainder with electrical service equipment may be equipped with only a 20-ampere, 125-volt receptacle.

Since most travel trailers and recreation vehicles built recently are equipped with 30-ampere receptacles, an acceptable arrangement is to install a power pedestal in the corner of four lots so that four different vehicles can utilize the same pedestal. Such an arrangement requires three 30-ampere receptacles and one

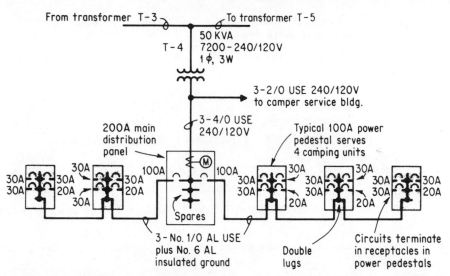

Figure 13–10   WIRING DIAGRAM SHOWING THE DISTRIBUTION SYSTEM OF A TRAILER PARK.

20-ampere receptacle to comply with Section 551–44 of the NEC. A wiring diagram showing the distribution system of a park electrical system serving 20 recreation vehicles lots is shown in Figure 13–10.

Electric service and feeders must be calculated on the basis of not less than 3600 watts per lot equipped with both 20-ampere and 30-ampere supply facilities and 2400 watts per lot equipped with only 20-ampere supply facilities. The demand factors set forth in Table 551–44 of the NEC are the minimum allowable factors that may be used in calculating load for service and feeders.

Example 13–1

Park area "A" has a capacity of 20 lots served by electricity; park "B" has 44. Find:

a   The diversity (demand) factor for area A

b   The diversity (demand) factor of area B

c   The total demand of area A

d   The total demand of area B

Solution

a   The diversity factor is 26 percent, read directly from Table 551–44 of the NEC.

b   The diversity factor is 24 percent, read directly from Table 551–44 of the NEC.

c   Since each lot is calculated on the basis of 3600 watts, the total demand is 20 × 3600 × 0.26 = 18,720 watts.

d   The total demand is 3600 × 44 × 0.24 = 38,016 watts (total demand)

QUESTIONS

Answer the following questions by filling in the blanks.

1   Article _____ of the NEC covers the electrical equipment and wiring methods for mobile home parks; Article _____ covers recreation vehicles and parks.

2   A unit that is towed by another vehicle and unfolds for use at a campsite is defined as a _____.

3   An electrical service must not be mounted _____ or _____ a mobile home.

4   Lighting loads are computed on the same basis as conventional residences; that is _____ watts per square foot.

5   The diversity allowed when sizing services for mobile homes after the first 3000 watts is _____ percent.

6   All equipment used outside the mobile home for electrical applications must be _____ proof.

7   Up to _____ power outlets may be utilized for a mobile home park service without needing a *main* disconnect.

8   Auxiliary gutters may contain conductors so long as their fill does not exceed _____ percent of the interior, cross-sectional area of the gutter.

9   A wire trough should not contain more than _____ conductors.

10   An _____ nipple is often desired for connecting safety switches, meter bases, etc., to wire troughs.

PROBLEMS

1   A mobile home is 40 feet long and 8 feet wide and has three portable appliance circuits; it also has a 300-watt oil furnace motor and 1200-watt water heater. Size the electric service for this mobile home as described in this chapter.

2   Refer to Table 551–44 of the NEC and find the total demand (in watts) for a travel trailer park containing 110 lots.

14

LOW VOLTAGE WIRING

Signal and communication systems are normally operated on low-voltage circuits, and these systems are usually installed similarly to conventional electrical circuits for light and power. NEC regulations governing low-voltage wiring are covered in Chapters 6 and 7.

A signal circuit, such as may be used for a burglar or fire alarm system, may be classified either as an open circuit or as a *closed* circuit. An open circuit system is one in which current flows only when a signal is being sent. In sending the signal, the N.O. (normally open) contacts are closed. On the other hand, current flows continuously in a closed-circuit system, except when the circuit is opened to cause signals to be sent.

All signal and communications sytems should be carefully planned, laid out, and installed to insure the best possible system. If working drawings are not available or if the original drawings are altered, sketches should be made as the system is installed to facilitate troubleshooting or additions to the system at a later date.

Wire sizes for the majority of low-voltage systems run from No. 22 AWG to No. 18 AWG. However, where larger-than-normal currents are required or when the distance between the outlets is long, it becomes necessary to use wire larger than specified in order to prevent excessive voltage drop. Just because you happen to be working with low voltage, do not hesitate to perform voltage drop calculations to determine the correct wire size for a given application.

14-1   BELL SYSTEMS

One of the simplest and most common low-voltage electric signal systems is the chime or door bell system. Such a system contains a low-voltage source, one or more pushbuttons, and a bell, buzzer, or chime.

Dry-cell batteries were once used for the low-voltage source; however, where alternating current is available, a transformer is now used for almost all low voltage systems. A step-down transformer is shown in Figure 14-1. The primary winding is connected to a 120-volt source, and the secondary terminals supply a low-voltage source of between 10 to 24 volts, depending on the application.

One pushbutton is sometimes used to activate the chime or bell, but it is more common to utilize two or more (one at each door), especially for residential applications. The set of chimes—located in a central location inside the house—normally has a code that identifies which button has been activated: the front door button may ring twice, the rear door once, and the side door three times. A wiring diagram for a two-note chime is shown in Figure 14-2. This diagram shows the chimes being controlled from two locations—the pushbutton at the front door sounds two notes when pushed whereas the button at the rear door sounds only one note when pushed.

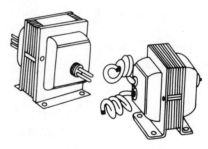

Figure 14-1   A STEP-DOWN TRANS FORMER OF THE TYPE COMMONLY USED FOR LOW-VOLTAGE SYSTEMS.

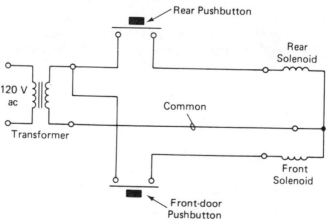

Figure 14-2   WIRING DIAGRAM SHOWING A TYPICAL TWO-NOTE CHIME CONTROLLED AT TWO LOCATIONS.

On systems with one signal unit—such as a vibrating buzzer or bell—the signal may be activated from any number of points by merely connecting the N.O. pushbuttons in parallel, as shown in Figure 14–3.

In apartment houses, private clubs, and similar occupancies, low-voltage systems are used as door openers and door bells for each apartment. A wiring diagram for such a system is shown in Figure 14–4.

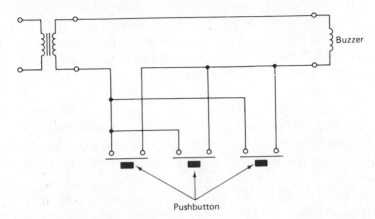

Figure 14–3   ON SYSTEMS WITH ONE SIGNAL UNIT, THE SIGNAL MAY BE ACTIVATED FROM ANY NUMBER OF POINTS BY CONNECTING THE PUSHBUTTONS IN PARALLEL.

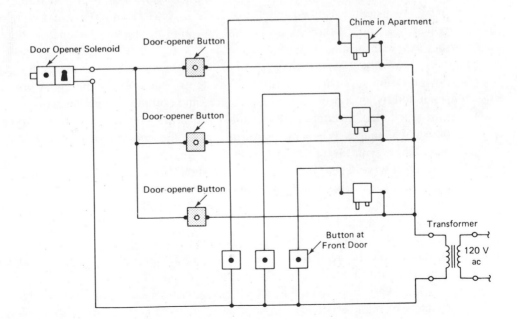

Figure 14–4   WIRING DIAGRAM OF A COMBINATION DOOR BELL AND DOOR OPENER SYSTEM.

Apartment House Intercom Systems

To provide apartment house tenants the security of a locked, building entrance door (excluding undesirable persons from the premises), many apartment building owners are installing apartment house intercom systems. With such a system, the caller must ring the apartment he is visiting and announce himself over the intercom system. The tenant, if satisfied with the identity of the caller, electrically opens the building entrance door.

A block wiring diagram of an apartment house intercom system is shown in Figure 14–5. Note that a transformer is connected to a 120-volt, a.c. supply. This transformer provides 16-volt alternating current to the amplifier (3087 Amplitone) unit for operation of the system. The amplifier is the heart of the intercom system. Most of the newer installations are solid state, capable of providing one watt of undistorted power within the range of normal voice frequencies. The power is carried from the transformer to the amplifier by a pair of No. 18 AWG conductors. Another pair of No. 18 AWG conductors runs from the amplifier to the door opener.

The units identified as "Suitefones" are equipped with three buttons: "Talk," "Listen," and "Door." Sometimes an extra button marked "Rear" is used to open the rear or service door.

Signal and communication wiring for chimes and apartment intercoms almost always consists of 18 AWG conductors. These conductors are normally enclosed in a two- or three-wire NM-sheathed cable and, in wood frame buildings, are usually installed much the same as NM (Romex) cable for light and power, i.e., pulled through drilled holes in joint or studs or fastened to the wood member with insulated staples.

Low-voltage wiring may not be installed in the same conduit with light and power systems, nor may the different systems occupy the same outlet box unless separated by an approved partition. Signal and communication systems are sometimes installed in conduit, especially where the conductor may be subject to physical damage. When raceways are used, the conduit should be installed in accordance with the tables in Chapter 9 of the NEC.

In all low-voltage systems, the wiring should be made-up in a connection box wherever several wires are brought together at one point. The wires should be fastened to connection strips, as shown in Figure 14–6. Each terminal should be marked for identification so that troubleshooting is simplified and the defective circuit can be readily disconnected without disturbing any of the other wires. Different-colored insulations should be used on the wires so that any one wire may be readily traced, as shown in Figure 14–5.

It is very important that all connections be made tight and secure. Loose contacts of high resistance can act like open circuits to the very low voltages of signaling systems. Where more than one wire has to be put under one screw or nut, separate washers should be used on top of each conductor.

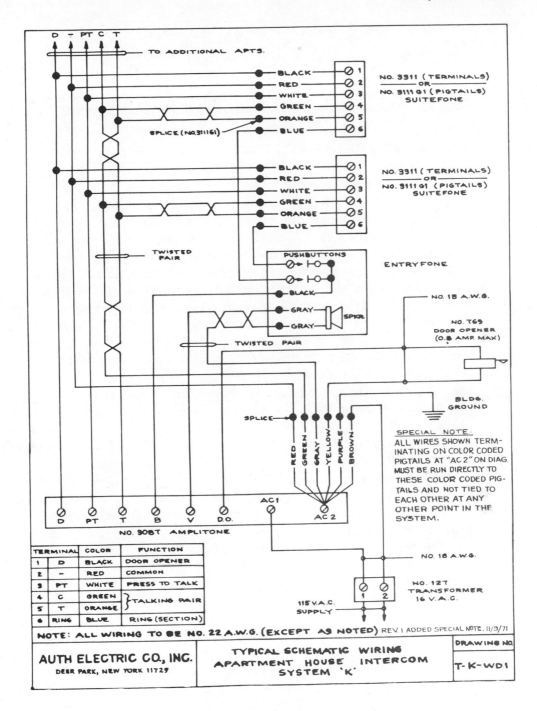

Figure 14–5   BLOCK WIRING DIAGRAM OF AN APARTMENT HOUSE INTER-
COM SYSTEM. (COURTESY, WEBSTER-AUTH ELECTRIC CO.)

Figure 14–6   WHERE SEVERAL LOW-VOLTAGE WIRES ARE BROUGHT TO-GETHER AT ONE POINT, TERMINAL STRIPS SHOULD BE USED.

14–2   ANNUNCIATORS

There are several types of nurses' call systems used in nursing homes and hospitals of all sizes. One type, manufactured by Webster-Auth, utilizes a combination of solid state and electromechanical components to accomplish their functions. This system is designed mainly for small hospitals and nursing homes.

The system consists of patients' call stations and lamp-type signal stations of various types. When a patient originates a call to the nurse, a steady lamp signal is lighted at various locations, depending on the requirements of the system, viz., at the bedside station (or stations) in his room, in the corridor over the room door, at the nurses' station, and in one or more duty rooms. In addition, a buzzer sounds at the nurses' station and in the duty rooms. Lamp signals are only extinguished when the nurse attends the patient at his bedside and presses the RESET button on the wall station. If the nurse requires emergency help when attending the patient, she can summon assistance by alternately pressing and releasing the CALL button. This action sounds the buzzer repeatedly at the nurses' station and in the duty rooms.

When the plug of any portable cord set is accidentally removed from its receptacle, all signals are energized in exactly the same manner as when the patient calls the nurse, except the buzzer sounds continuously. Various interchangeable cord sets are available for patients' use: standard cord sets for normal use, pressure pad for the aged and incapacitated, and pull switch for oxygen-tent patients. All plug into the same receptacles on the patient's bedside station.

Emergency-type stations in toilets, baths, solariums, etc., enable the patient to summon assistance by pulling a pendant cord attached to a switch. This flashes a lamp signal on all associated signal stations and repeats a buzzer signal intermittently and insistently at the nurses' station and in the duty rooms. These audible and visible signals can be cancelled only by throwing the switch on the originating station to the OFF position.

The circuitry employed in this system makes possible the use of standard, momentary contact pushbutton cord sets and standard, two-point, jack-type receptacles. This lowers the cost of the original equipment and the cost of replacement components.

A typical layout of such a system is shown in Figure 14–7. Note the symbols and nurses call wire legend. The first digit of the number circling a wire run indicates the number of No. 22 AWG wires in the run, and the second number indicates the number of No. 18 AWG wires in the run. Thus, the number "6–5" indicates that the run or conduit contains six No. 22 AWG conductors and five No. 18 AWG conductors. The catalog numbers correspond to those components manufactured by Webster-Auth.

14–3 INTERCOMMUNICATION SYSTEM

Before an intercom system is installed within a building, the overall system should be carefully and thoroughly planned. The most desirable location for and height of the speakers should be selected; speakers and controls should be located so they will not be blocked by a piece of furniture at a later date. When a system is installed in an existing finished building, the routing of the cables should be given careful consideration in order to cut down on cutting and patching of the room finishes.

The first consideration should be given to the various components available for the intercom system, of which there are over a dozen, including AM and FM radio, TV audio, record changer, three types of tape decks, intercom (including talk–listen door speakers), door chimes, and security/fire alarm system—all included in one central system. Most units on the market are designed to fit between standard partition studs (14 inches on centers) and are therefore easily installed in either or existing partitions provided rough lumber was not used for the studs. In this latter case, a half inch or so will have to be chiseled from the two studs enclosing the unit to make the unit fit.

During the rough-in stage, a 120-volt, a.c. power supply, an antenna lead-in cable for the FM system, and speaker and control wiring have to be provided. In music-only systems, two pairs of No. 18 AWG wire will suffice for all low-voltage wiring; for stereo/intercom systems, seven pairs of No. 18 AWG wire will be needed from the master station to each remote station.

Speakers may be either ceiling or wall mounted, but all speakers exposed to the outdoors in either walls or soffit must be weather resistant. Stereo speakers should be installed within the same wall approximately two-thirds of the wall length apart, but never closer than four feet or farther than 15 feet apart. Each set of speakers should be controlled by a remote volume control location in or near the area where the speakers are located.

Begin the installation of an intercom/music system by selecting the preferred locations for the master station, remote stations, door speakers, remote control units, and alarm devices, if they are used. For maximum operating convenience, install the master and all inside stations 4-1/2 feet from finished floor to the bottom of the unit. Also make certain that the master unit is located at least 4 inches from adjacent walls, cabinets or countertops.

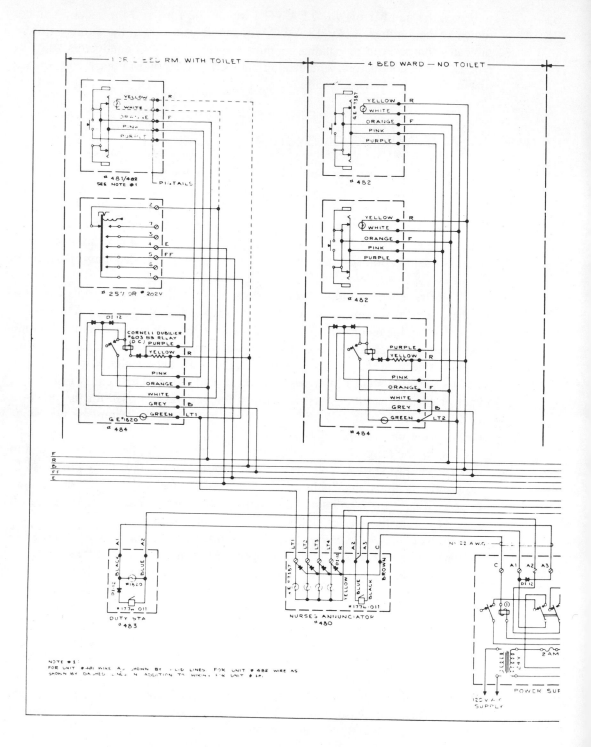

Figure 14–7 TYPICAL LAYOUT OF A NURSES' CALL SYSTEM.

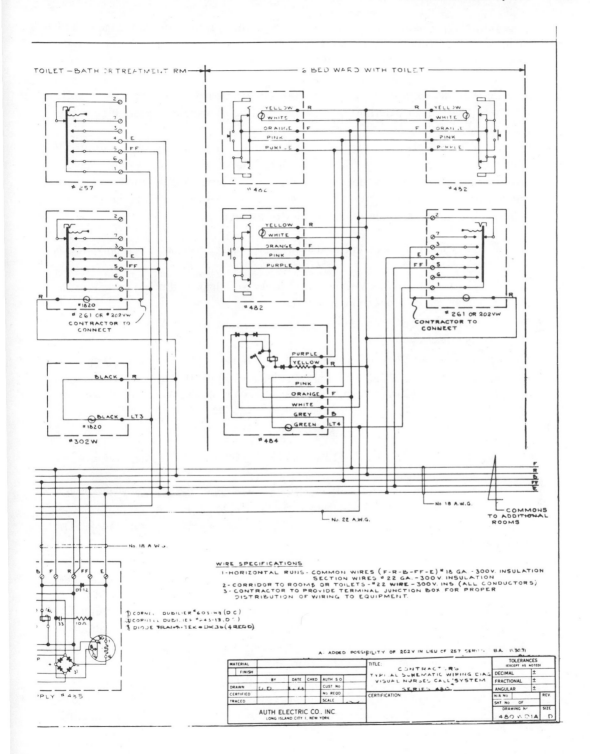

Figure 14–7   (Concluded)

Whenever possible, try to locate the inside units on interior walls that are free of insulation and other obstructions. To prevent feedback or interference, never mount speakers or any other of the controls or devices back to back or on a common wall between rooms. Door speakers, of course, have to be located on an exterior wall adjacent to the entrance door or in a nearby porch ceiling or soffit overhang. Again, all exterior components must be of weather-resistant construction.

An AM/FM antenna should be provided for the radio if an existing TV antenna cannot be utilized. Most self-contained stereo music systems have a built-in antenna already connected to the set at the factory. However, if the built-in antenna does not give the customer satisfactory results, try an alternate antenna arrangement, such as one of the following:

1   Fully extend the 120-volt line cord to ensure that the built-in antenna is providing its peak performance.

2   Disconnect the built-in antenna and connect a T-shaped (dipole) antenna to the 300-ohm terminals on the back of the set.

3   Disconnect the built-in antenna and connect an indoor TV/FM antenna to the 300-ohm terminals.

4   Disconnect the built-in antenna and connect an outdoor TV/FM antenna. If the leads are 300 ohms or a 75-ohm coaxial cable with a 75 to 300 ohm transformer on the end is used, connect it to the 300-ohm terminals; if you prefer to use only the 75-ohm cable with split connectors, connect it to the 75-ohm terminals.

14-4   TV AND FM ANTENNA LEAD-IN DISTRIBUTION SYSTEMS

Cable TV systems are used mostly in towns and cities. Rural areas still depend on TV antenna systems designed for far-fringe areas up to 200 miles away from TV stations. A diagram of such a system is shown in Figure 14–8. The components are described below.

Antenna head   There are several antenna heads for VHF, UHF, and FM reception that take care of most reception situations ranging from local through far-fringe areas. Most range in size from 9 elements, good for a range of about 30 miles, to 67 (or more) elements for distances to over 225 miles. The terrain, of course, has much to do with the reception. For example, if the terrain was flat, the reception range is greater on flat terrain than on hilly or mountainous terrain.

Most modern antenna heads are preassembled with fold-out elements to ease installation. All come with complete instructions for assembling and connecting.

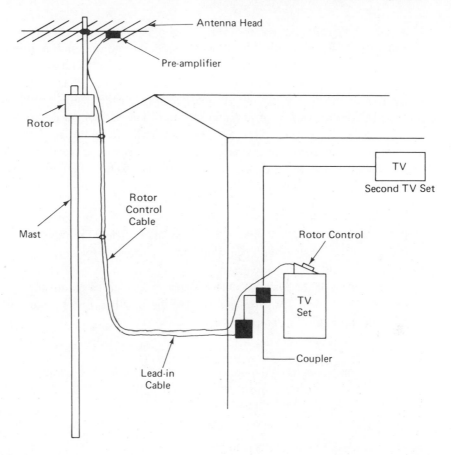

Figure 14–8   DIAGRAM OF A TV AND FM ANTENNA LEAD-IN DISTRIBUTION SYSTEM.

Pre-amplifier Pre-amps are designed primarily for use in fringe areas to boost the signal and most come in two units: one part mounts on the antenna mast, and the other is placed near the TV set. In general, the mast-mounted pre-amplifier overcomes downlead losses and rejects interference while matching the roof antenna to the TV or FM set under all atmospheric and weather conditions. An indoor-mounted, a.c. power supply plugs into a conventional 120-volt wall outlet and sends current to the mast-mounted pre-amplifier. The combination of the two amplifies channel signals 2 to 83 plus FM reception. If more than one TV set is to be used on the system, power-boosters are manufactured for use with four or more sets.

Rotator   The antenna rotator system consists of a motor-operated rotator mounted on the mast, a control unit inside the home, and a four-conductor power/control cable connecting the two. The inside unit connects to a 120-volt power source for proper operation. Once the rotator system has been mounted and properly connected, set the dial compass setting on the inside control unit to the direction that you want your antenna to point. The rotator on the mast points the antenna to that direction and turns off automatically when the direction is reached.

Another semiautomatic type of control works in a very similar fashion, except that a control bar is depressed for the motor to turn the antenna and must be held down until the best picture is seen.

Masts   Mast kits for TV antennas come in a variety of types from chimney mounts, through roof mounts, ground mounts secured by brackets connected to the house, to self-guying towers.

The height of the antenna mast is very important to obtain the best possible reception. If the local TV shop cannot give the best height for a particular area, a test should be made to determine this.

Lead-in cable   Basically, there are four types of lead-in cables.

1   Twin-lead, 300-ohm, ribbon-type cable

2   Twin-lead, 300-ohm, foam-insulated cable

3   Twin-lead, 300-ohm, shielded permohm cable

4   Seventy-five-ohm, shielded coaxial cable.

The ribbon-type, 300-ohm, lead-in cable is usually considered to produce the strongest signal under adverse environmental conditions in low-interference sub-urban areas. However, problems occur when the cable passes near metallic objects and where high-interference conditions exist, such as in metropolitan areas.

Twin-lead, foam-insulated cable is also a flat-type cable but is totally encased in and surrounded by polyethylene foam, backed up with an outer polyethylene jacket. This type of cable offers a high resistance to ultra-violet rays, oil, fumes, moisture, salt air, and abrasion.

Another type of 300-ohm cable is the shielded type. Designed for 82-channel, color-TV reception, this type of cable combines the strong signal strength of twin-lead cable with the clean signal of shielded coaxial cable. The shield helps to eliminate ignition and other interference caused by line pick up.

The 75-ohm cable, although more costly than most other types, provides greater efficiency by minimizing interference and is highly water resistant and is easier to install. This type of cable can be installed anywhere, even over metal objects. However, since antenna output and TV set input are at 300 ohms, a step-down (300—75-ohm) and a step-up (75—300-ohm) transformer must be used for the signal transfer.

Couplers When two or more TV sets are fed from a single antenna system, an all-channel, multiset coupler must be used to divide the signals evenly among the sets and to eliminate interset interference. The most common types for use around the home are designed for two or four sets.

Antenna signal splitter A signal splitter separates all-channel signal into individual VHF/UHF/FM signals for TV set input. The impact device mounts on the back of the TV set; the all-channel lead-in wire is connected to the correct terminals on the unit; a separate wire for every individual signal (VHF, UHF, and FM) connects to three other sets of terminals on the back. Both VHF/UHF splitters and VHF/UHF/FM splitters are common.

TV Outlets

The first steps in installing TV outlets are to locate the outlets in various areas of the building, to plan the routing of 75-ohm coaxial cable from the antenna to the first outlet, and to loop from outlet to outlet. For single-floor houses with an accessible crawl space or basement under the first floor, all the cable can be fed from underneath without too much cutting, drilling, and patching. For example, locate the outlets in the various rooms, cut an opening for a conventional plastic outlet box, locate the point directly under the outlet box opening and in the partition, and drill a 1/2-inch hole through the wood floor, making certain that the drilling stays well within the boundaries of the wall partition in which the outlet box opening is cut. Repeat this procedure at each of the outlet locations before pulling the 75-ohm cable from the antenna head to each of the outlet locations. After the coaxial cable is looped from outlet to outlet and is pulled inside of each of the outlet boxes with about 12 inches of lead-in wire hanging out of the box, install conventional 75-ohm to 300-ohm TV outlet receptacles with covers.

14-5 BURGLAR ALARMS

All burglar alarm systems have three common functions: detection, control, and annunciation or alarm signaling. Most detectors incorporate switches or relays that operate because of entry, movement, pressure, infrared beam interruption, etc. The control senses detector operation with a relay and produces an output that may operate a bell, siren, silent alarm, etc. The controls also frequently contain ON/OFF switches, test meters, time delays, power supplies, standby batteries, and terminals for tying the system together. The control output usually provides power on alarm to operate bells, sirens, etc., or switch contacts for silent alarms, such as automatic telephone dialers.

A pictorial diagram of a typical burglar and fire alarm system is shown in Figure 14–9. This particular system can accommodate at least 20 burglar detectors, 3 smoke detectors, and an unlimited number of thermostat fire detectors that can be added at any time.

The burglar alarm circuit permits use of magnetic switches, switch mats, and ultrasonics for intruder detection. Two zones are incorporated into the unit under discussion: one zone is used for an outside perimeter guard whereas the second connects interior motion detectors and floor mats.

Conventional burglar alarm systems usually employ contacts at all openings (doors and windows). However, the contents of some buildings are so valuable that burglars may try to enter through walls or ceilings. The most economical method of detecting this type of forced entry is to use vibration detectors. These detectors are designed to initiate an alarm from vibration and to detect burglary attempts through walls and ceilings by impact attacks with hammer, crow bar, or any heavy object. The detectors should be installed on walls, such as brick, hollow tile, concrete and plaster, about 4 or 5 feet from the floor and from 2-1/2 to 6 feet apart, depending on wall thickness and construction. In Figure 14–10, for example, on an 8 inch brick wall the detectors are from 4-1/2 to 5 feet from the floor and at 6 foot centers. Tests have shown this to be the proper spacing; however, further tests on a specific installation may indicate greater or lesser spacing.

Detectors are mounted vertically on angle brackets fastened to 1 inch × 2 inch wood strips, which are normally painted to match the area finish. Detector units and wiring of burglar and fire alarm systems are installed like any other type of low-voltage signal system, i.e., locating the outlets, furnishing a power supply, and connecting the components with No. 18 or 22 AWG cables. Some different types of wire and cable for signal wiring are shown in Figure 14–11.

Most of the work in installing an alarm system is "simple" wiring among detectors, controls, remote controls, bells, etc. Certain systems, such as those used in warehouses, use exposed wiring or easily concealed wires above false ceilings. These may suffice, but concealed wiring is much more secure against tampering or defect by either insiders or outsiders. Finally, appearance is always important: exposed wiring is seldom decorative.

Alarm systems using wiring for interconnection are usually far superior to others since they are more secure and will last longer. Other types, such as radios and house wiring, are not supervised and depend on relatively unreliable electronic circuits. Most closed systems use No. 22–2 or 24–2 AWG wire and are color coded for identification. Size 18–2 is normally adequate for connecting bells or sirens to controls, if the run is 40 feet or less. Many electricians prefer to use No. 16–2 or even 14–2 cable to be sure.

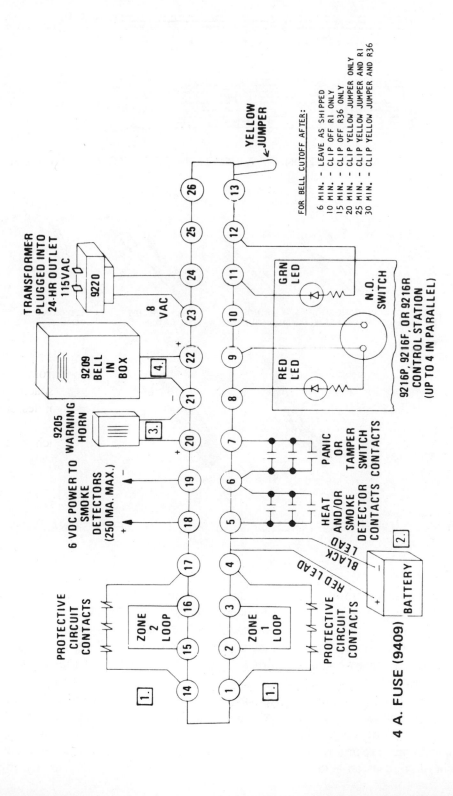

Figure 14–9 PICTORIAL DIAGRAM OF A TYPICAL BURGLAR AND FIRE ALARM SYSTEM.
(COURTESY, MOUNTAIN WEST ALARM)

239

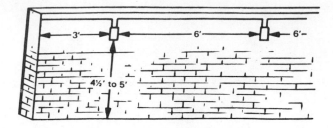

FIGURE 14–10 INSTALLATION OF VIBRATION DETECTORS ON 8-INCH THICK BRICK WALL.

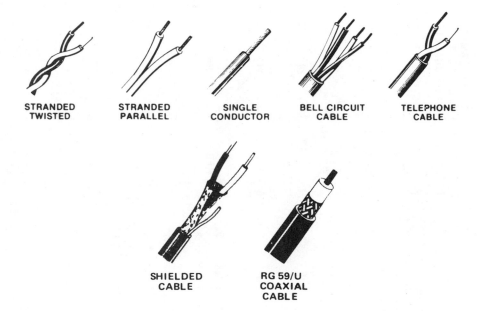

Figure 14–11 DIFFERENT TYPES OF WIRE AND CABLE FOR SIGNAL WIRING.

Alarm systems are often classified according to the means of sounding the alarm or alerting someone who can respond effectively to a break-in or fire. A local alarm, for example, includes a detector or control unit to provide on-off, testing, power, etc., and a signaling device, such as a sounder light, to indicate the alarm locally. The disadvantage of a local alarm is that the intruder knows that he has been detected.

Another type is the silent alarm. This system utilizes a detector, a control, and a means of sending an alarm by wire to a remote location where action can be initiated. The alarm control output may be sent by leased telephone line to a central station where professionals can follow up with police action. Another kind of silent alarm uses an automatic telephone dialer that utilizes phone service to contact several parties such as the police, fire department, alarm company, owner, or foreman.

14-6   CLOCK SYSTEMS

The primary control unit of a master time and program system is commonly called the "master clock." It is normally a wall-mounted panel assembly installed in the office of a responsible official in schools. It is actually a master controller and performs two major functions:

1   It is wired to a central source of unswitched power and operates all other clocks in the system. Isolated power interruptions in a building do not affect remote secondary clocks. At fixed schedule periods, which may be hourly or every twelve hours, the master controller transmits synchronizing signals to all secondary clocks. This ensures that the time reading of all system clocks remains identical to that of the master.

2   The mechanism includes a unique, multicircuit control feature called a programmer that permits the master controller to transmit additional signal impulses to external circuits, which connect to devices such as bells, horns, chimes, or buzzers. These are referred to as program signals. Program circuits are also utilized for ON-OFF control of building utilities.

Program signal control is established and maintained by use of continuous-run, prepunched "memory tapes." Each circuit is programmed independently and uses a principle indentical to those used in tape control of communication transmission networks and similar to that of punch-card data processing.

In installing such a system, it is necessary to obtain certain details concerning the system for the proper roughing-in of outlet boxes, conduit, cable, etc. If the working drawings do not contain sufficient information, rough-in details are usually available from the system's manufacturer.

A typical block diagram of an electric clock system is shown in Figure 14–12. Note that a 120-volt, 60-Hz power supply feeds the master clock and the control rectifier. Control wiring from the master clock includes four-wire and two-wire circuits to the control rectifier. Two-wire circuits are also looped from the master clock to secondary clocks. The wiring is performed in the usual manner, either with cable or raceway, depending on the building structure through which the circuits must pass.

14-7   REMOTE-CONTROL SWITCHING

To supply the demands for greater convenience and flexibility of lighting control for office buildings, institutions, commercial and industrial buildings, and residences, low-voltage switching systems were designed. In this type of system, relays are used to perform the actual switching of the current. These relays are in

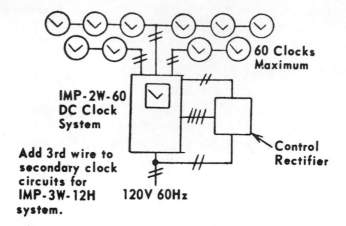

IMP-2W-60 DC Clock System

60 Clocks Maximum

Control Rectifier

Add 3rd wire to secondary clock circuits for IMP-3W-12H system.

120V 60Hz

Figure 14–12   TYPICAL BLOCK DIAGRAM OF AN ELECTRICAL CLOCK SYSTEM.

turn controlled by small switches that operating at a low voltage, permiting the use of wiring similar to that used for residential door chimes. Because these relays only require a momentary impulse to change from OFF to ON or ON to OFF, the control switches are momentary-contact switches, meaning that the low-voltage current, of small magnitude, flows only for the length of time that the switch is pressed. These momentary-contact switches provide several important advantages:

1   An unlimited number can be wired in parallel since none are actually "in" the circuit except during the short period that they are being pressed.

2   One type of switch performs the function of a single-pole switch, a three-way switch, and a four-way switch with no complicated wiring.

3   Switches can be added to existing circuits at any time without changing the wiring.

In a low-voltage system, the load-carrying conductors run directly to the relay, while the control switches use light, inexpensive conductors that make the many loops to all the switches. Since the switching is done at a point remote from the control switches, the system is appropriately called a *remote-control wiring system*. If switch points are arranged close together so that a dial-type switch can sweep many contacts in a fraction of a second, a semimaster control of many circuits is accomplished; user may control, say, 25 different circuits by merely pressing one control switch.

A wiring diagram of a typical remote-control switching circuit is shown in Figure 14–13. The branch circuit feeding the load is installed in the usual manner. However, instead of running directly to the load or to a conventional wall switch,

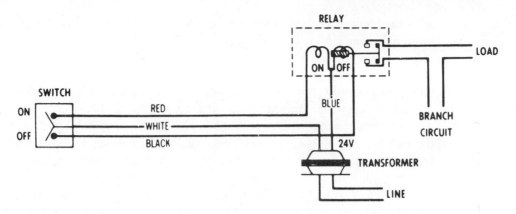

Figure 14–13 WIRING DIAGRAM OF A TYPICAL, REMOTE-CONTROL
SWITCHING CIRCUIT.

the circuit runs to the relay. The split-coil relay shown in the diagram permits positive control for ON and OFF. It can be located near the load or be installed in centrally located distribution panel boxes, depending on the application. Because no power flows through the control circuits and low voltage is used for all switch and relay wiring, it is possible to place the controls at a great distance from the source or load.

The low-voltage power is obtained from a 120/240-volt transformer. One lead from the transformer is connected to the center top of the relay, and the other lead connects to the common terminal on the switch. When the switch is pressed to the ON position, the circuit is completed through the ON coil in the relay. This sets up a magnetic field and pulls the plunger to the left (in the illustration), closing the contacts; this in turn completes the circuit to the load. When the OFF button is pressed, the circuit is completed through the OFF coil, drawing the plunger to the right. This opens the contacts and breaks the circuit feeding the load.

In most low-voltage switching applications, a central location for installing the relays for the control of a single floor or confined area provides maximum flexibility for changes in control, ease in servicing, and shorter runs of low-voltage wires for automated control systems. When centralized relay installations are considered, special relay and component cabinets offer the advantages of an integrated design for the installation of relays, low-voltage power supply, motor-master control units, and all necessary bus bars for ease in wiring.

When relays are located in plenum, it is advisable to use one or more transformer/rectifier combinations mounted in the plenum area to eliminate running control wires back to the distribution box. Some installations use small, flush-mounted boxes in hallways for the zone relay. This eliminates the need to enter the plenum when the switch control is changed to meet these requirements or the requirements of new room partition locations.

When relays are located in a ceiling plenum, flexible metallic raceways (BX, etc.) are generally used to connect the relays to the lighting fixtures. These are either premanently connected to the relay boxes or are mounted in the relay boxes through locking caps and matching locking receptacles.

The component cabinets hold several remote-control relays, mounted through 1/2-inch KOs (knockouts) provided in the cabinet walls. Mounting holes facilitate installing the transformer/rectifier assembly and one or two motor-master control units.

Built-in terminal strips are provided for the common (white) conductors, and the line-voltage connections to the relays are made with clamp-type terminals, four terminals on each relay. A typical component cabinet showing internal wiring is depicted in Figure 14–14.

Most wiring for low-voltage, remote-control switching should be of No. 20 AWG, thermoplastic-insulated type with two, three, or four conductors as required and be rated for 30-volt service. Where possible, all wiring should be run concealed, without conduit, in ceiling spaces and between walls and partitions. Wiring run in masonry or exposed to the weather should be run in conduit. The installation methods are similar to those described for other types of low-voltage circuits.

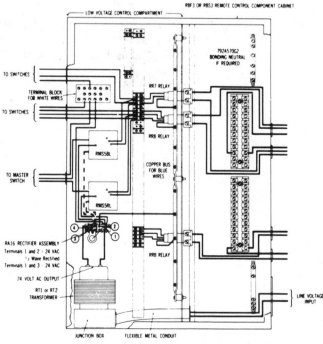

Figure 14–14 TYPICAL COMPONENT CABINET SHOWING INTERNAL WIRING.

QUESTIONS

Answer the following questions by filling in the blanks.

1   Wire sizes for most low-voltage systems will range from No. _____ AWG to No. _____ AWG.

2   In most low-voltage systems, the low-voltage source is obtained by using a _____ connected to a 120-volt source.

3   On low-voltage systems with one signal unit, the signal may be activated from any number of points by connecting the pushbuttons in _____ with each other.

4   Signal and communication wiring for chimes and apartment intercoms almost always consist of No. _____ AWG conductors.

5   One means of identifying the different low-voltage conductors is to use _____ insulations on the wires.

6   When residential intercommunication systems with many components are installed, a 120-volt, a.c. power supply should be provided during the rough-in stage as well as _____ lead-in cable for the FM system and TV.

7   When speakers for a stereo/intercom system are selected, all speakers exposed to the outdoors must be _____ resistant.

8   In general, stereo speakers should never be installed closer than _____ feet or farther than _____ feet apart.

9   When a TV/FM antenna head is selected for distances of approximately 200 miles from the closest TV station, an antenna head with _____ or more elements should be selected.

10   The type of TV antenna lead-in cable that provides the greatest efficiency by minimizing interference is the _____-ohm _____ cable.

11   The three common functions of burglar alarm systems are _____, _____, and _____.

12   The primary control unit of a master time and program system is the _____.

13   The basic components of a remote-control switching system consist of a _____ to feed the load, a _____ to provide a low-voltage source, a _____ to do the actual switching, one or more _____ to control the switching, and _____ to connect the components.

14   Any number of remote-control switches may be used to control a single relay; therefore, a cable consisting of _____ wires will be required between switches in order to connect them together.

15   Chapters _____ and _____ of the NEC deal extensively with signal and communication systems.

15

EMERGENCY
ELECTRICAL SYSTEMS

The fundamental need for a standby emergency source of electric power during a power source outage becomes evident when the reason for such an installation is considered. Normal supply of electric power may be interrupted at any time and for many reasons. Most of the time, the cause is nothing more than a mechanical breakdown that lasts only a few minutes. However, power outages are often caused and accompanied by other dangers to our physical well being. Storms, hurricanes, floods, and explosions not only strike down poles and power lines but leave a trail of property damage and personal injury as well. It is precisely during such critical emergency periods that vital public electric services are needed in order to operate:

1   Emergency lighting

2   Power for elevators

3   Police and fire departments

4   Radio and TV stations

5   Life-saving equipment in hospitals

Some states require that all public buildings (schools, offices, etc.) have a standby electric power source. Banks rely heavily on standby electric power to ensure that their alarm systems operate at all times—under any conditions. Some industrial and commercial firms connect the majority of their electric lighting, office machines, and PBX telephone switchboard to a standby electric power source: either battery banks or gasoline driven electric generators (see figure 15–1).

Emergency generators are used extensively where ventilating or refrigerating apparatus is necessary at all times. Hotels are usually required by law to have some source of emergency power to operate lights, elevators, and other services essential to the safety and convenience of guests.

In high-rise office buildings, standby electric generating sets are used to provide power to operate a certain number of emergency lights on each floor, elevators, sump pumps, and fire pumps. In fact, standby electric power systems are becoming increasingly essential in all public buildings as well as in dwelling units since more and more people depend on electric power for work and everyday living.

Articles 700 and 750 of the NEC give regulations governing emergency and standby generating plants from the standpoint of design, installation, and testing. It is suggested that these articles be read thoroughly before designing or installing an emergency electric system.

15–1 EMERGENCY GENERATORS

When an emergency generating set is considered, there are several factors involved. Two of the most important points to consider initially are:

1   Availability of the various fuels in the area where the generator is installed.

2   Local regulations governing the storage and usage of gasoline and gaseous fuels.

One of the most important factors in selecting a standby generator is its size or capacity. A careful study should be made to determine exactly what degree of service it is required to maintain during a power outage; that is, the kind of equipment (d.c., a.c., single phase, or polyphase) and the total load in kVA or watts of each.

It is advisable in selecting the generator to adopt a long-range view since it often becomes necessary to add more equipment to the load in future years. It may be desirable to provide full protection at the outset since the higher cost of a larger unit is offset by special circuit wiring costs necessary for selected protection. As needs grow, the set may still be adequate restricting if protection is restricted to essential equipment. Of course, the decision of whether to provide full or selected, restricted protection is generally made by the particular purpose for which the unit is needed. A hospital, for example, requires full protection for all essential equipment whereas it may be desired to maintain emergency lighting only in halls and corridors in a residential building.

Another selection factor is the altitude of the building, and the effect it may have on the rating of the standby set. If the point of installation is at a higher altitude than the maximum set by the manufacturer, the unit capacity is derated. It is generally derated 4 percent per 1000 feet above sea level. Manufacturers can furnish this data on request.

Before the engine can be selected, careful consideration must be given to the type of fuel. Often, the type of fuel is one of the most important considerations in engine selection. Fuel options are gasoline, LP gas, natural gas, and diesel.

Air and liquid cooling systems are usually required, particularly for large-capacity engines. Liquid cooling may be either a radiator or a fresh-water cooling system. Generally, the determining factor between air and water cooling is the size of the unit. Air cooling is confined to units below 15 kW. In these smaller sizes, it is possible to force air over the engine and generator at a sufficient rate to cool it properly and thus to take advantage of lower equipment and maintenance costs. Larger units are usually equipped with a closed liquid system for maximum engine-cooling protection.

Regardless of capacity or cooling system, the location, size, and (ambient) temperature of the room must be considered. City-water cooling, remote radiators, and heat exchangers are also available for special applications.

15–2 TRANSFER SWITCHES

There are two types of transfer switches, manual and automatic. The *manual* transfer switch is operated by a person on duty either at the generator or at one or more remote stations. The generator is started by a manual switch, and the load is transferred from one source to another by a hand-operated, double-throw switch.

An *automatic* transfer switch senses the power outage and automatically starts the generator. It transfers the load automatically, without requiring the attention of an operator. With automatic transfer switches, the duration of a power outage can be limited to less than 10 seconds. An automatic, load transfer switch is almost a necessity for applications where uninterrupted power is of prime inportance, e.g., a hospital, where equipment is remotely located, or where health or safety is at stake.

The operating features of an Onan transfer switch are shown in Figure 15–1. The *electrical interlocks* (contacts 4 and 5) provide positive, electrically guided action from NORMAL to STANDBY position. The *mechanical interlock* bar shown in the drawings acts as an added safety feature against both sources (power line and generator), supplying the load simultaneously. Both interlocks insure that the switch has only one position at any one time.

When NORMAL line power is supplying the load (Figure 15–1a), the line side contacts (1) are closed, allowing current transfer from the normal supply to the load. The latch holds the mechanical interlock bar and line coil armature in the UP position, mechanically locking the line contacts closed. During normal power operation, the coil disconnect switch (3) is in an OPEN position, which deenergizes the line coil, helping to eliminate a.c. vibration and hum. The generator coil electrical interlock switch (4) is also OPEN during normal operation; this prevents the generator contacts (2) from closing.

When the normal line power fails, a start-stop relay in the control circuit starts the generator, and current from the generator energizes the *latch coil*. This releases the mechanical interlock bar, opening line side contacts (1) and closing the generator coil electrical interlock switch (4) (Figure 15–1b). With this switch closed, current flows from the emergency generator through the generator coil, closing the generator side contacts (2).

Once the generator is supplying the load, the line coil electrical interlock switch (5) opens, preventing the line contacts (1) from closing. At the same time, the mechanical interlock bar locks the line coil armature DOWN, providing double protection against the line contacts closing and both sources being connected ot the load. Latch coil disconnects switch (6) opens, deenergizing the latch coil (Figure 15–1c).

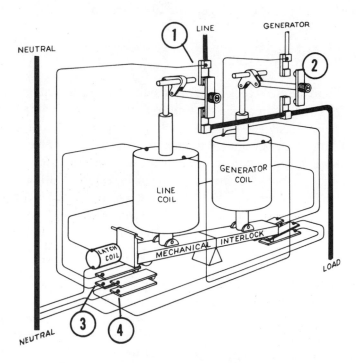

Figure 15–1a DRAWING SHOWING AUTOMATIC TRANSFER SWITCH IN NORMAL POSITION.

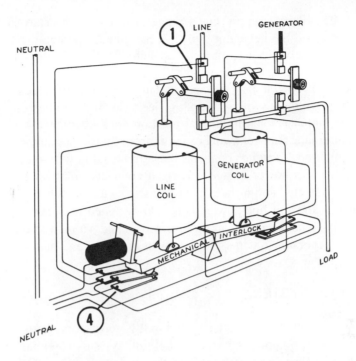

Figure 15–1b   TRANSFER SWITCH DURING TRANSFER ACTION.

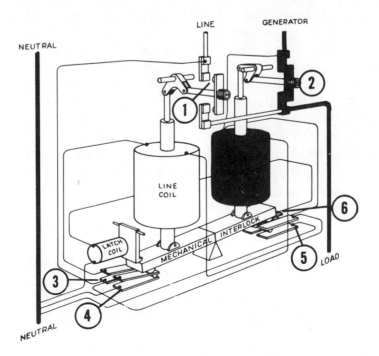

Figure 15–1c   AUTOMATIC TRANSFER SWITCH IN STANDBY POSITION.

When normal supply line voltage is restored, the start-stop relay stops the generator, and the generator coil is deenergized, opening the generator side contacts (2). The electrical interlock switch (5) closes, allowing current to flow through the coil disconnect switch (3), energizing the line coil and closing line contacts (1). The latch again mechanically locks the line side contacts closed, and the coil disconnect switch (3) opens, deenergizing the line coil. At this point, the transfer switch is back in normal position once again.

Automatic transfer switches range from 30 to 4000 amperes in capacity. The automatic transfer switch connects the commecial power source to the load during operation. If a power outage occurs, it connects emergency power to the load. Properly selected and installed, the transfer switch will handle the full generating set output and full line power. It is also designed to break the currents it passes and to pass peak current many times its rated value without damage. Typical standby generator systems are shown in Figures 15–2 and 15–3.

15–3 BATTERY SYSTEMS

Instead of electromechanical standby generators, battery systems are frequently used to supply emergency lighting systems essential to safety, life, and property in the event of failure of the normal lighting. Generally, emergency lighting is designed into a building for one purpose, to provide *minimal illumination* for personnel safety and building evacuation. Minimal illumination is defined as sufficient light to demark, at least by silhouette, any obstructions or safety hazards in the direct travel path. The Life Safety Code requires that ''every exit and the necessary ways of exit access thereto shall be illuminated . . . continuously during the time that the conditions of occupancy require that the means of egress be available for use. . . . The floors of exits and of ways to exit access shall be illuminated at all points such as angles and intersections of corridors and passageways, stairways, landings of stairs, and exit doors. Emergency lighting facilities shall be arranged to maintain the specified degree of illumination in the event of failure of the normal lighting for a period of 1-1/2 hours.''

Based upon this code, some guidelines for the placement of battery-operated emergency lights are presented below.

Hallways, corridors, etc.   Emergency lighting units should be mounted at every intersection or change of direction of hallways, lobbies, and entrances to fire stairwells. In straight corridors, units should be ceiling mounted on 45 foot centers. See floor plan in Figure 15–4. Corridors up to 100-feet long can be effectively lighted with a single lighting unit; this unit should be wall-mounted and aimed down the corridor opposite the exit door when coupled with a self-powered exit light above the exit door (Figure 15–5).

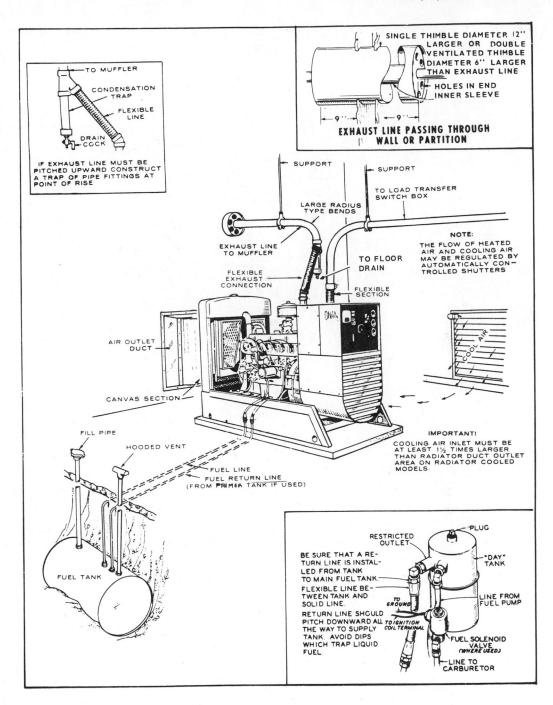

TO MUFFLER
CONDENSATION TRAP
FLEXIBLE LINE
DRAIN COCK

IF EXHAUST LINE MUST BE PITCHED UPWARD CONSTRUCT A TRAP OF PIPE FITTINGS AT POINT OF RISE

SINGLE THIMBLE DIAMETER 12" LARGER OR DOUBLE VENTILATED THIMBLE DIAMETER 6" LARGER THAN EXHAUST LINE
HOLES IN END INNER SLEEVE
9" 9"
EXHAUST LINE PASSING THROUGH WALL OR PARTITION

SUPPORT
SUPPORT
TO LOAD TRANSFER SWITCH BOX
LARGE RADIUS TYPE BENDS
EXHAUST LINE TO MUFFLER
FLEXIBLE EXHAUST CONNECTION
TO FLOOR DRAIN
FLEXIBLE SECTION

NOTE:
THE FLOW OF HEATED AIR AND COOLING AIR MAY BE REGULATED BY AUTOMATICALLY CONTROLLED SHUTTERS

AIR OUTLET DUCT
ONAN
COOL AIR

CANVAS SECTION

FILL PIPE
HOODED VENT
FUEL LINE
FUEL RETURN LINE (FROM PRIMER TANK IF USED)

IMPORTANT!
COOLING AIR INLET MUST BE AT LEAST 1½ TIMES LARGER THAN RADIATOR DUCT OUTLET AREA ON RADIATOR COOLED MODELS.

FUEL TANK

RESTRICTED OUTLET
PLUG
"DAY" TANK

BE SURE THAT A RETURN LINE IS INSTALLED FROM TANK TO MAIN FUEL TANK.
FLEXIBLE LINE BETWEEN TANK AND SOLID LINE.
RETURN LINE SHOULD PITCH DOWNWARD ALL THE WAY TO SUPPLY TANK. AVOID DIPS WHICH TRAP LIQUID FUEL.

TO GROUND
TO IGNITION COIL TERMINAL
LINE FROM FUEL PUMP
FUEL SOLENOID VALVE (WHERE USED)
LINE TO CARBURETOR

Figure 15-2 TYPICAL GASOLINE INSTALLATION OF AN EMERGENCY GENERATOR PLANT.

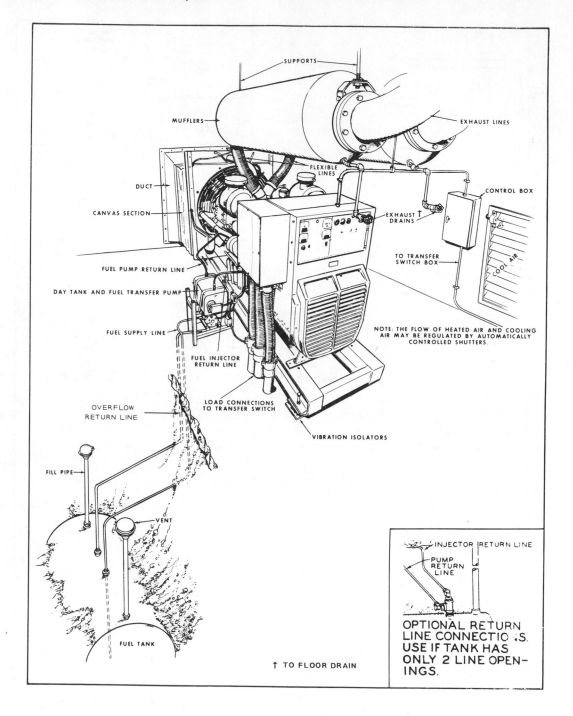

SUPPORTS

MUFFLERS

EXHAUST LINES

FLEXIBLE
LINES

DUCT

CANVAS SECTION

CONTROL BOX

EXHAUST
DRAINS

TO TRANSFER
SWITCH BOX

COOL AIR

FUEL PUMP RETURN LINE

DAY TANK AND FUEL TRANSFER PUMP

FUEL SUPPLY LINE

FUEL INJECTOR
RETURN LINE

NOTE: THE FLOW OF HEATED AIR AND COOLING
AIR MAY BE REGULATED BY AUTOMATICALLY
CONTROLLED SHUTTERS.

OVERFLOW
RETURN LINE

LOAD CONNECTIONS
TO TRANSFER SWITCH

VIBRATION ISOLATORS

FILL PIPE

VENT

INJECTOR RETURN LINE

PUMP
RETURN
LINE

OPTIONAL RETURN
LINE CONNECTIONS.
USE IF TANK HAS
ONLY 2 LINE OPEN-
INGS.

FUEL TANK

† TO FLOOR DRAIN

Figure 15-3 TYPICAL DIESEL INSTALLATION OF AN EMERGENCY
GENERATOR PLANT.

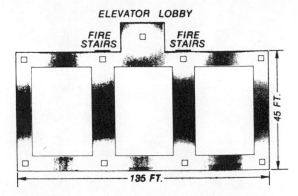

Figure 15–4 EMERGENCY LIGHTING LAYOUT ON ONE FLOOR FOR HALL-WAYS AND CORRIDORS IN A LARGE OFFICE BUILDING.

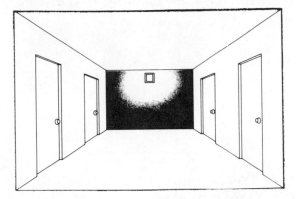

Figure 15–5 VIEW OF EMERGENCY LIGHTING DOWN A 100-FOOT LONG HALL FROM EGRESS DOOR.

Stairwells One emergency lighting unit adequately illuminates a stairwell when either wall- or ceiling-mounted at every landing. One unit per floor with one additional unit located over the point of exit from the stairwell provides adequate illumination (Figure 15–6).

Area lighting Emergency lighting in large open spaces with high ceilings, such as auditoriums, cafeterias, factories, warehouses, meeting halls, and gymnasiums, may be illuminated with battery-powered units mounted on 75-foot centers (Figure 15–7) to allow evacuation of the area and to prevent panic.

Security lighting In areas where injury or property loss might readily occur during a power failure, a single battery-powered unit can provide excellent security protection. Typical locations include cash registers; bank teller stations; bank vault entrances; high-value merchandise displays; machines that operate from an

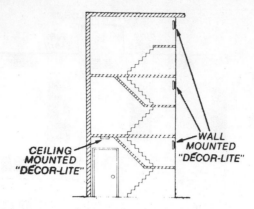

Figure 15–6 EMERGENCY LIGHTING LAYOUT FOR A STAIRWELL. THE FIXTURES LOCATED AT THE LANDINGS CAN EITHER BE WALL OR CEILING MOUNTING.

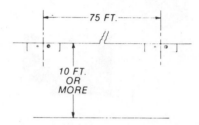

Figure 15–7 TYPICAL, EMERGENCY-LIGHTING MOUNTING FOR LARGE OPEN SPACES WITH HIGH CEILINGS.

independent power source or will continue to move for a short period after power loss; electric service equipment locations such as switchgear, generators, and control panels; security and watchman stations; emergency communications stations; fire extinguishers, hoses, and hydrants; life-saving equipment; and dangerous obstructions or shaftways.

Restaurants, bars, and lounges   The lighting levels in these facilities are normally low. One battery-powered lighting unit mounted on the wall over the point of exit of an unobstructed room provides sufficient lighting for up to 5000 square feet. For a larger area, the fixtures should be mounted 40 feet apart (on the average) on centers.

Convalescent and nursing homes   Studies indicate that the eyes of the senior citizens are unable to adjust rapidly to sudden decreases in light levels. Therefore, it is extremely important to take this into consideration when lighting units for emergency lighting systems are located in this type of occupancy. When battery-powered lighting units are mounted on no more than 30-foot centers in large areas and hallways and are center-ceiling-mounted in smaller rooms, the necessary level of lighting to allow the aged to evacuate the building will be maintained during power failure.

Theaters and lecture halls   Subdued lighting normally utilized in this type of facility reduces the lighting level required for emergency lighting. One, average, battery-powered lighting unit mounted on the ceiling suffices for areas up to 5000 square feet. When these units are supplemented by exit lights with DOWN lights at each point of exit (with additional units in the lobby, stairs, rest rooms, and coat check areas), emergency lighting requirements are adequately met.

Sizing Wire for Battery-Operated Systems

Circuit runs feeding remote, battery-operated equipment must be of sufficient wire size to maintain a proper operating voltage to all lamps. The maximum allowable voltage drop is 5 percent. To select the proper wire size, use any form of the following equation, depending on which factor is unknown:

$$L = \frac{CM \times VD}{I \times 22} \tag{15-1a}$$

$$CM = \frac{I \times 22 \times L}{VD} \tag{16-1b}$$

$$VD = \frac{I \times 22 \times L}{CM} \tag{15-1c}$$

$$I = \frac{CM \times VD}{L \times 22} \tag{15-1d}$$

where CM = wire area in circular mils, I = load in amperes, L = length of circuits from battery to load in feet, and VD = maximum allowable voltage drop (usually 5 percent of the supply voltage) in volts.

The above equations may be used to calculate the wire area in circular mills that is required to carry the specified load at a predetermined and allowable voltage drop. Table 15–1 identifies the necessary wire size from the corresponding circular mil area calculated.

Table 15–1   WIRE SIZE TABLE FOR LOW-VOLTAGE EMERGENCY WIRING

Wire size (AWG)	Capacity in amperes	Ohms per 1000 feet	CM size
12	20	1.586	6,530
10	25	0.9972	10,380
8	35	0.6271	16,510
6	50	0.3944	26,250
4	70	0.2480	41,740

Example 15–1

A 12-volt, emergency-lighting system runs a distance of 75 feet from source to load, carrying a current of 8 amperes, maximum. Use a maximum allowable voltage drop of 5 percent of the selected voltage to calculate:

a   The maximum voltage drop in the lines.

b   The circular mil area of the wire to be used.

c   The commercial AWG gauge wire size to be used.

d   Explain why a No. 12 wire capable of carrying 20 amperes is not used.

Solution

a $VD = (12 \text{ volts}) = 0.05 = 0.6 \text{ volt}$

b $CM = \dfrac{I \times 22 \times L}{VD} = \dfrac{8 \times 22 \times 75}{0.6} = 22,000 \text{ CM}$

c   From Table 15–1, use a No. 6 AWG conductor, having an area of 26,250 CM (the next largest size).

d   Although the current is only 8 amperes, the length of the run is too long and the resistance per foot of the No. 12 wire is too high. Such a choice would produce a voltage drop in excess of 5 percent of the supply voltage.

Example 15–2

Assume that a No. 6 wire is actually used in Example 15–1 for the 75-foot run, carrying a maximum current of 8 amperes. Calculate:

a   The actual voltage drop in volts.

b   The actual percent voltage drop of the supply voltage.

c   Explain why the voltage drop is actually below 5 percent.

Solution

a $VD = \dfrac{I \times 22 \times L}{CM} = \dfrac{8 \times 22 \times 75}{26,250} = 0.503 \text{ volts}$

b   Percent voltage drop $= \dfrac{0.503 \text{ volts}}{12 \text{ volts}} \times 100 = 4.2 \text{ percent}$

c   By using a larger CM area (26,500) than was claculated in Example 15–1 (22,000), we obtain the voltage drop that is lower and is within the 5 percent allowable value.

The smallest permissible wire size for emergency lighting equipment under Article 720–4 of the NEC for systems under 50 volts, regardless of the calculated CM area, is No. 12 AWG. Table 15–2 gives the proper wire size for various lengths of circuit runs at 6, 12, 32, and 115 volts. It may be used for quick reference, but the basic equations (15–1) should be understood before the Table is used.

Table 15–2 WIRE SELECTION TABLE FOR RUNS OF VARYING LENGTH AT DIFFERENT VOLTAGES

6-VOLT
WIRING DISTANCES IN FEET

WIRE SIZE	WATTS 13	18	28	44	50	100	150	200	250
#12	45'	30'	18'	12'	11'	6'	4'	3'	2'
#10	74'	49'	30'	20'	18'	9'	6'	5'	4'
#8	118'	78'	47'	31'	28'	14'	10'	7'	6'
#6	185'	125'	75'	50'	45'	23'	15'	13'	9'

12-VOLT
WIRING DISTANCES IN FEET

WIRE SIZE	WATTS 13	18	25	28	32	44	50	100	150	200	250	300	320
#12	166'	111'	81'	73'	62'	47'	41'	20'	13'				
#10	265'	190'	136'	121'	104'	79'	69'	34'	23'	17'	13'		
#8	415'	300'	215'	194'	166'	124'	108'	54'	36'	27'	21'	18'	16'
#6	660'	465'	340'	305'	260'	198'	172'	86'	58'	42'	34'	28'	25'

32-VOLT
WIRING DISTANCES IN FEET

WIRE SIZE	WATTS 13	18	25	28	32	44	50	100	150	200	250	300	320	400	500
#12	1200'	905'	550'	500'	490'	350'	310'	155'	108'	79'	61'	52'	49'	40'	30'
#10		1475'	1025'	880'	780'	560'	500'	250'	168'	125'	98'	83'	78'	63'	50'
#8				1400'	1250'	890'	800'	400'	265'	200'	155'	133'	125'	100'	75'
#6						1400'	1250'	625'	425'	325'	250'	210'	200'	163'	125'

115-VOLT
WIRING DISTANCES IN FEET

WIRE SIZE	WATTS 13 AMPS .113	18 .156	25 .217	28 .243	32 .278	44 .382	50 .434	100 .869	150 1.30	200 1.73	250 2.17	300 2.60	320 2.78
#12	14,964'	10,839'	7,792'	6,958'	6,082'	4,486'	3,896'	1,945'	1,300'	977'	720'	650'	608'
#10	23,787'	17,230'	12,367'	11,661'	9,669'	7,636'	6,193'	3,093'	2,067'	1,553'	1,238'	1,033'	966'
#8	37,810'	27,410'	19,705'	17,596'	15,361'	11,193'	9,852'	4,820'	3,289'	2,471'	1,970'	1,644'	1,538'
#6	60,159'	43,570'	31,327'	27,975'	24,455'	17,795'	15,663'	7,822'	5,229'	3,929'	3,132'	2,614'	2,445'
#4	95,663'	69,294'	49,815'	44,485'	38,884'	28,298'	24,907'	12,493'	8,315'	6,248'	4,981'	4,157'	3,888'

QUESTIONS

Answer the following questions by filling in the blanks.

1   Two of the first points to consider when an emergency generating set is selected are
_____ and _____.

2   Fuel options for most standby emergency generator sets are _____,
_____, _____, and _____.

3   The cooling of generator sets is handled by either _____ or
_____ cooling systems.

4   The two basic types of transfer switches are _____ and _____.

5   Automatic transfer switches range in size from _____ amperes to _____ amperes.

6   Most transfer switches are designed to break the currents it passes and to pass
_____ current many times its rated value without damage.

7   Emergency lighting facilities are arranged to maintain a specified degree of illumination in
the event of failure of the _____ lighting for a period of _____ hours.

8   Three typical security lighting applications are _____,
_____, and _____.

9   The maximum allowable voltage drop for battery-fed circuits feeding emergency lighting
is _____ percent.

10   Article _____ of the NEC states certain provisions covering the wire size to be used for
emergency lighting equipment under 50 volts.

PROBLEMS

1   Use the equations in the text to determine what size wire should be used for a 32-volt
emergency lighting system drawing a current of 11 amperes for a distance of 60 feet, the
voltage drop should not exceed 5 percent.

2   Use Table 15–1 to determine what size wire is required to feed two-50-watt lamps
connected to a 12-volt circuit 50 feet from the battery.

3   In Problem 2, what wire size would be required if the distance were increased to 75 feet?

16

EXTERIOR WIRING

Outside electrical work is generally recognized as all electrical construction that is installed outside of the individual building lines, proper. The various categories of this wiring include all types of underground work outside the building lines; wood-pole, overhead-line construction; steel-tower, overhead-line construction; substation and switchyard construction; and overhead trolley systems. Much of this work is handled by specialty contractors employing trained linemen who work only on high-voltage, power distribution systems. However, the material in this chapter deals only with outside circuits and feeders (under 600 volts) for recreation lighting, flood lighting, power distribution between buildings, and the like.

Basically, there are three methods of installing outdoor wiring: in metallic or PVC (plastic) conduit using weather-tight fittings; underground wiring using raceways buried in the ground or direct-burial cable; and overhead wiring using single-conductor or "messenger-type" cable suspended on poles or between buildings. All of these are satisfactory for outdoor wiring, but there are certain cases where one type is preferred over another.

16–1 OVERHEAD DISTRIBUTION

Overhead systems range from small, single-pole farm (and residential) distribution lines to large, steel-tower transmission lines. The higher voltages are usually classified as transmission lines whereas the lower voltages are classified as distribution lines. We deal only with second distribution lines of a normal voltage of 600 volts and under in this text.

Overall lines should be constructed in accordance with the basic requirements of the NEC and local ordinances. Most fall into one of the following three categories:

1   Individual, bare or insulated, conductors supported by pin-type insulators mounted on insulator pins, which are mounted on wooden crossarms, steel brackets, etc.

2   Insulated wires supported by spool-type insulators incorporated in steel racks mounted on the sides of poles.

3   Insulated, multiconductor aerial (messenger) cables either self-supporting or supported from steel messenger cables.

Weather-proof, insulated cables are recommended for all nongrounded conductors for secondary distribution lines under 600 volts (see NEC Section 310–5).

Wooden poles are the most common means of supporting overhead lines and are generally classified according to type of wood, height, circumference, straightness, knots, and preservative treatment. The butts of poles are generally treated at the pole supply yard with a wood preservative to a point 1-1/2 feet above ground level, or in some cases, for their full length.

The tops of poles are usually sawed off at an angle. When wooden crossarms are mounted on the pole, a notch or "gain" is cut in the side of the pole to provide a flat bearing surface. The height of poles is governed by the size of conductors, number of circuits carried, weight of equipment to be mounted on the pole, change of direction of the line, etc.

For secondary distribution systems, overhead lines using pin-type or light-weight groups of suspension insulators or insulator racks are usually constructed on individual poles. The distance between poles varies, depending on size and strength of the conductors, difference in grade elevation of adjacent poles, change of direction of the line, and operating or service conditions. For secondary distribution systems, the distance between poles is seldom over 150 feet.

Wood poles are held in an upright position by placing them in holes in the ground. The depth of the holes depends on the height of the poles and degrees of change in the direction of the line. Poles set in soft ground will obviously have to be set deeper than those set in hard rocky ground. The depth of pole holes is usually not less than three feet. The diameter of the hole should not be any larger than is necessary to allow the use of tampering bars to tamp the backfilled dirt properly. Concrete is sometimes used to backfill around poles in poor ground.

Pole hardware includes all of the various types and sizes of machine bolts, carriage bolts, lag screws, washers, eye nuts, plain eye bolts, thimble eye bolts, pole steps, clevisses, insulator pins, brackets, dead-end conductor clamps, steel guy cable, protective guy plates, guy hooks, guy rods, anchors, guy guards, and bolted guy clamps.

Screw type insulators are the most commonly used insulators for secondary distribution systems. Spool-type insulators mounted on a metal rock or used with metal clevisses are also quite common.

Pole-line structures are subjected to strains resulting from the pull of conductors, change in direction of the line, weight of equipment, and wind. These strains must be counteracted in some manner to prevent breaking off, upsetting, swaying, or disaligning the pole structure and undue sagging of the conductor. This is accomplished principally by the use of dead-end or lateral "down" guys attached to some form of anchor placed in the ground, as shown in Figure 16–1.

A guy consists of a stranded steel guy cable, the ends of which are passed through the eyes of either regular or thimble eye nuts or bolts fastened to the pole structure at the upper end and the eye of an anchor rod at the lower end and through the holes in the insulator (if used) and secured with bolted guy clamps. The free end of the cable is secured to the running section of the cable with a serving of iron wire or special clips.

Individual conductors used for overhead power lines range from solid No. 8 copper to large, bare, stranded cables. In the larger sizes, various combinations of stranded copper or aluminum and steel cable are used where the copper or aluminum conductor does not have sufficient tensile strength to support its own weight or where the added weight of ice is frequently a problem. Hard drawn or medium-hard drawn conductor is used rather than soft drawn because of greater tensile strength. Weather-proof, covered conductor is used on the lower-voltage distribution circuits.

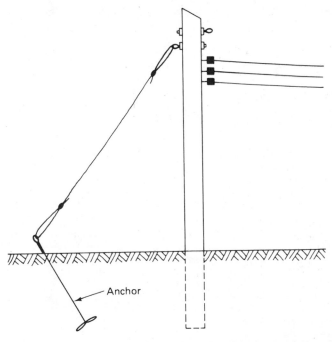

Anchor

Figure 16–1 STRAINS ON WOOD POLES MAY BE COUNTERACTED BY ATTACHING GUY WIRES TO SOME FORM OF ANCHOR PLACED IN THE GROUND.

16–2   UNDERGROUND DISTRIBUTION

In areas where the soil is not too rocky, underground outside secondary distribution systems are the preferred method (NEC Articles 230–30 to 230–33). Underground distribution systems have the following advantages over other types of outdoor wiring systems.

1   Underground lines are not subject to damage as are overhead lines: overhead lines are sometimes struck by trucks passing beneath them and often break from heavy ice loads during winter months. Repairs and maintenance on underground lines are much easier than overhead lines, in most cases.

2   Underground lines cannot be seen and therefore present a better appearance. A lot of overhead lines give a "cluttered" look to even the most well-kept areas.

3   Underground wiring, when cable is used, is often less expensive to install than other types of outdoor wiring.

Where soil conditions permit, direct-burial cable installations are the most economical and are often more convenient to install. The installation of the cables in a direct-burial system is either performed by laying them in a trench or by plowing them under with a special cable plow. Cables installed in a trench are often laid on a layer of sand (Figure 16–2) and covered with sand and a one-inch (or thicker) creosoted plank to protect the cable and workers in the area from injury, which might occur if a digging tool pierced the cable. A warning ribbon is sometimes laid in the trench, just under the ground surface, to warn diggers that electrical cable is buried beneath the ribbon.

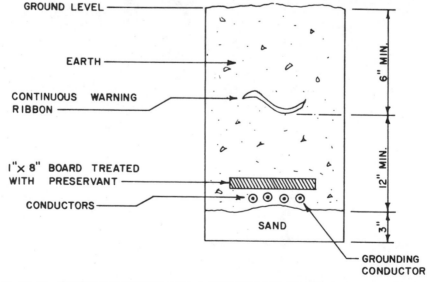

Figure 16–2   SECTION THROUGH A TRENCH SHOWING DETAILS OF THE INSTALLATION PROCEDURES.

Underground raceway systems (NEC Article 230–32) consist of junction boxes and connecting runs of one or more rigid, PVC, or fiber ducts placed in trenches. The conduits provide a routing for the conductors while the junction boxes are provided to allow for the splicing continuous conductor runs, the junction of lateral runs, and the installation of operating equipment.

In nearly all instances when multiraceway systems are laid, they are installed by the built-up method. This involves laying the conduits in an excavated trench on either a concrete base or on precast, concrete separators. The full height of the conduit formation is erected before the concrete, which encloses and envelopes it, is placed. If a concrete base is used, it should be mixed comparatively dry and be placed and rammed in the trench to form an evenly graded layer of the required thickness, to the height of the grade stakes previously established. When precase separators are substituted for the concrete base, they should be placed across the trench at the required intervals and leveled so that they form the grade for the first duct layer and provide a space of about three inches between the bottom of the trench and the conduits. (See NEC Articles 347–2 and 347–3).

The laying of the conduit is begun on the concrete base before it has received its initial set or on the precast concrete separators if that method of construction at the bottom of the trench is used. The precast concrete separators are placed near each end of the ducts, horizontally over the first layer and vertically between the individual ducts. As the laying of duct proceeds, the individual conduits and layers are tied together and to each other with pieces of heavy wire. A rigid skeleton structure is progressively built to the height of the full formation by this method.

Pouring the concrete follows the duct laying and is carried to a height at least three inches above the top tier of the duct or conduit structure. Figure 16–3 shows several cross sections of underground duct or conduit systems.

Electricians working on electrical systems for building construction seldom become too involved in complete underground electrical systems, but they occasionally have short runs of conduit between buildings to install. Four-inch, galvanized or PVC (plastic) conduits are frequently used for the short runs and are positioned about 6 inches apart on centers and about 30 inches below the ground surface. They are sloped to manholes or buildings for drainage.

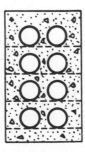

Figure 16–3   CROSS SECTION OF AN UNDERGROUND CONDUIT SYSTEM.

Manholes are needed to avoid excessive lengths of cable, to avoid curves in the conduit lines, and for intersections. Most modern manholes are constructed of concrete, prefabricated at a plant, and delivered to the job site. They may, however, be constructed at the job site out of brick or concrete. They are usually spaced 300 to 400 feet apart.

Large cables are normally ordered in lengths to fit the conduit sections, with an adequate allowance for slack and splicing. Smaller cables are usually shipped on standard reels and are cut to the required lengths on the job. To pull the cable, a cable grip is attached to one end of the cable, and a pulling rope or cable is used. Cables may be pulled by hand although the use of a cable-pulling machine or a wrench is often desirable. In any case, wire-pulling grease should be used to facilitate the pull.

16–3 USE OF SPECIAL POWER TOOLS AND EQUIPMENT FOR INSTALLING EXTERIOR WIRING

Regardless of the purpose of exterior wiring—outdoor lighting, power, transmission, communications, etc.—there are several special tools available that facilitate the installation of such systems. An electrician should become familiar with these tools and learn how to use each to its best advantage.

Trenches for underground electrical distribution systems are either hand dug or machine dug. For short runs, hand digging usually suffices. But the use of *compressed-air jack hammer* with appropriate drill, spade, and tamping accessories greatly reduces labor and fatigue.

In areas that have no adverse ground conditions or other utilities in line with or crossing the new trench run, the use of *trenching* or *ditching machines* is more economical than hand digging. Self-propelled mechanical trenching equipment ranges from small hand-tractor type of trenchers (Figure 16–4) to large riding trenchers (Figure 16–5). When the extent of the electrician's employer's work does not warrant the purchase of a trencher, such equipment can normally be obtained on a rental basis.

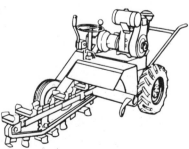

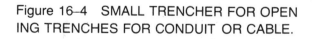

Figure 16–4 SMALL TRENCHER FOR OPENING TRENCHES FOR CONDUIT OR CABLE.

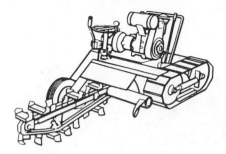

Figure 16–5   RIDING TRENCHER FOR OPENING TRENCHES OR DITCHES.

To install cable and conduit under roadways and sidewalks while avoiding damage to the surface of either, special *boring machines* are used to drill or push conduit under the walk or roadway. Such machines require only a narrow starting trench and terminal sump hole, making restoration of the area to its original condition less demanding. Drill heads range in size from 1-1/4 to 2 inches for the initial ''cutting'' pass and from 2 to 3-1/2 inches for the return or ''reaming'' pass.

Rotational torque for a boring machine is usually furnished by the machine itself. The thrust is supplied by an up-front pulling action of the drill head as it cuts its way into the soil, pulling both the sections of drill steel and machine behind it. Once through to the sump hole, the drill head is removed and a reamer head is attached. Flexible tubing, electrical cable, telephone cable, or conduit may be attached to the swivel on the reamer. As the reamer pulls itself back through the soil, it carries the tubing, cable, or conduit from the sump hole to the starting trench, and installation is completed. Connection of services and minimal clean up are all that remain to what used to be a mammoth task of repairing cut streets and backfilling and tamping large areas of disturbed earth.

When drilling under sidewalks or roadways is not practical, a suitable channel can normally be cut with a *concrete saw* (Figure 16–6). Two cuts are made with the saw where the conduit or cable will run. This can be done with little or no damage to the surrounding pavement. The gravel and dirt is removed to accept the conduit and cable, and the cut then needs only minor patching.

When several post holes need to be dug for wood poles for an outside distribution system, a *powered post-hole digger* (Figure 16–7) can speed the operation tremendously. If the contractor does not own one, he can usually rent one by the day, week, or month.

Other special tools include an electronic metal locator for locating underground cable and conduit, an underground fault locator for locating a break or ground fault in buried cable, and a power wire-puller for pulling long runs of heavy cable through underground raceways.

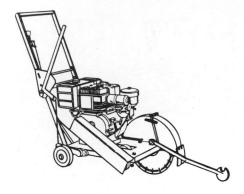

Figure 16–6 WHEN DRILLING UNDER SIDEWALKS OR ROADWAYS IS NOT PRACTICAL, A SUITABLE CHANNEL CAN NORMALLY BE CUT WITH A CONCRETE SAW.

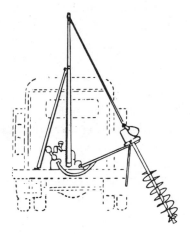

Figure 16–7 POWERED-POST HOLE DIGGER.

16–4 GFCIS ON OUTSIDE RESIDENTIAL CIRCUITS

In all one-family houses, each dwelling unit in a 2-family house, each apartment in an apartment house, and each dwelling unit in a condominium, ground fault circuit interruption (GFCI) must be provided in certain areas. In general, GFCI protection is required for all 120-volt, single-phase, 15- and 20-ampere receptacles installed in bathrooms and garages and all outdoor power outlets. The requirement for GFCI protection in garages was added in the 1978 NEC because homeowners use outdoor appliances plugged into garage receptacles. Protection may be provided either by a GFCI circuit-breaker (Figure 16–8) that protects the whole circuit and any receptacle connected to it or by the receptacle itself being a GFCI type that incorporates the components that give it the necessary tripping capacity on low-level ground faults.

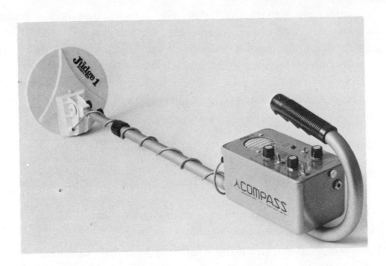

Figure 16–8 ELECTRONIC METAL DETECTOR.

Section 210–8 of the NEC requires that GFCI protection be provided on all 120-volt, single-phase, 15- and 20-ampere receptacles installed outdoors at dwelling units where there is direct, grade-level access to the dwelling unit and to the receptacles. "Direct grade level" may lend itself to many varied interpretations, but receptacles installed on balconies, porches, or other raised parts of the building are generally not required to have GFCI protection, provided there is no grade-level (ground) access to the receptacles. Such protection is not needed where the receptacle cannot readily be used by someone plugging in a tool or appliance and making contact with earth, masonry walk, or other surface or grade.

Ground-fault circuit interruptors (GFCIs, as shown in Figure 16–9) must be used to provide personnel protection for all 120-volt, 15- and 20-ampere recepta-

Figure 16–9 GFCI CIRCUIT BREAKER.

cle outlets on construction sites. GFCI protection is not required on 240-volt receptacles, three-phase receptacles, or receptacles rated over 20 amperes although it would be good practice to use GFCI even on these circuits.

It is recommended that the electrician keep abreast of all changes in the NEC and consult the latest NEC edition frequently as a guide when installing GFCI protection or any other phase of an electrical installation.

QUESTIONS

Answer the following questions by filling in the blanks.

1   Where soil conditions permit, the electrician may use _____ cable instead of installing a raceway system.

2   The most common means of supporting overhead electrical lines are _____.

3   The smallest size conductor normally used for overhead power lines is No. _____ AWG copper.

4   The distance between supports on secondary overhead distribution systems is seldom over _____ feet.

5   The installation of cables in a direct-burial system is performed either by laying them in a trench (either hand dug or mechanically dug) or by plowing them under with a special _____.

6   One method of protecting direct-burial cable is to lay a _____ on top of them.

7   Manholes might be needed in underground distribution systems to avoid _____ and _____ and for _____.

8   Where there is direct, ground-level access to an outside receptacle (installed on a dwelling unit), Section 210–8 of the NEC requires that _____ protection be provided.

17

TESTING ELECTRICAL INSTALLATIONS

In order to maintain and troubleshoot existing electrical systems, electrical workers should know and apply modern test techniques and have a good understanding of basic testing instruments.

When you use any testing instrument (or meter), always *consider your personal safety first*. Know the voltage levels and shock hazards related to all equipment to be tested and make certain that the instrument used has been tested and calibrated: this should be done at least once a year. To prevent damage to the instrument, select a range (on meters with different ranges) that insures less than full scale deflection of the needle. A mid-scale (or higher) deflection of the needle usually provides the most accurate reading.

17–1   VOLT-OHM-AMMETERS

The combination volt-ohm-ammeter is probably the electrician's most useful testing instrument for electrical systems under 600 volts. Many of these instruments are very compact and can easily be carried in a leather pouch attached to the electrician's belt for immediate use.

One type of volt-ohm-ammeter is shown in Figure 17–1. Instruments of this type are commercially available in current ranges from 6 to over 1000 amperes and voltages from 0 to 600 volts. A separate battery attachment is supplied for use of the ohmmeter scale built into the instrument.

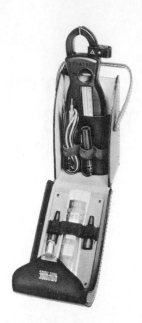

Figure 17–1   COMPACT VOLT—OHM—AMMETER.

Make sure that the battery attachment case is removed from the instrument when taking current or voltage readings. If the battery attachment is not removed, incorrect voltage or current readings will be obtained. The battery attachment is to be inserted into the instrument only when it is used as an ohmmeter.

To take current readings, release the pointer on the scale by moving the pointer lock button to the left. Turn the scale selector knob until the highest current range appears in the scale window. Press the trigger button to open the jaws before encircling *one* of the conductors under test with the transformer jaws. Never encircle two or more conductors, as shown in Figure 17–2a; only encircle one conductor as shown in Figure 17–2b. Release finger pressure on the trigger slowly to allow the jaws to close about the conductor and keep an eye on the scale

Figure 17–2   WRONG WAY TO MAKE A CURRENT READING WITH TRANSFORMER JAWS.

Figure 17–2b   RIGHT WAY TO MAKE CURRENT READING; NOTE THAT THE TRANSFORMER JAWS ENCIRCLES ONLY ONE CONDUCTOR.

while doing so. If the pointer jumps abruptly to the upper range of the scale prior to closing the jaws, completely, the current is probably too high for the scale used. Should this happen, release the jaws immediately and use either a higher scale or a "rating-busting" attachment, which will be explained later. If the pointer deflects normally, close the jaws completely and take the reading from the scale.

A reading below half scale indicates that an adjustment is necessary for a more accurate reading. There are two ways to do this. You can increase the accuracy by looping the conductor two or more times around the transformer jaws and then dividing the reading by the number of turns. If the instrument has a lower scale adjustment, it is easier to set the scale at the next lowest range. For example, if the 100-ampere scale is used and the pointer is below 40 amperes, as shown in Figure 17–3a, set the rotary scale selector to the next lower current range, and the reading will be in the upper half of the scale, as shown in Figure 17–3b. When very low current readings are encountered, another attachment (Figure 17–4) is available for increasing the current measuring sensitivity of the instrument 5 to 10 times. Therefore, the 0 to 6 ampere range becomes either 0 to 1.2 or 0 to 0.6 amperes.

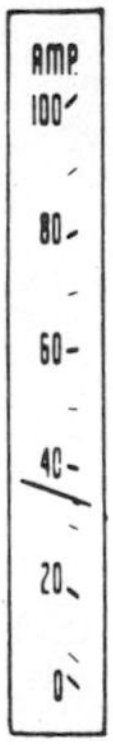

Figure 17–3a   WHEN A 100-AMPERE SCALE IS USED, THE POINTER IS IN THE LOWER HALF OF THE SCALE.

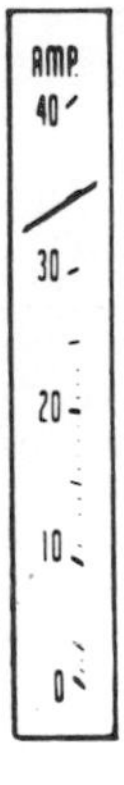

Figure 17–3b   WHEN A 40-AMPERE SCALE IS USED, THE POINTER IS IN UPPER HALF OF SCALE, GIVING A MORE ACCURATE READING.

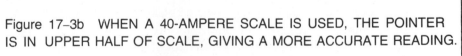

Energizer Model A-45L

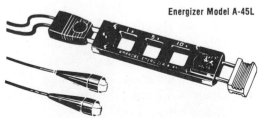

Figure 17–4   AN ADAPTOR FOR TAKING VERY SMALL CURRENT READINGS ON A STANDARD AMMETER.

Instruments with current surge indication capability may be used to measure motor starting currents. The procedure is as follows.

1   If the starting current (locked rotor) is not shown on motor nameplate, estimate the current 5 to 6 times full-load (nameplate) current.

2   Set the instrument to the appropriate current range and adjust the pointer to zero, using black, zero-adjust screw, if necessary, on instrument.

3   Turn the red, surge-adjustment screw counterclockwise to move the pointer upscale. Set to a value approximately 5 percent below estimated current. Do not try to set pointer above 95 percent of full-scale value of the range, e.g., on 100-ampere range, do not set above 95 amperes.

4   Turn off the motor.

5   Clamp the instrument jaws around one motor lead. Observe the pointer and turn on the motor.

6   Adjustment is correct when the pointer shows only a slight movement (less than one division) upscale. Read the current value to which pointer moves. If the pointer moves upscale more than one division or does not move at all, readjust pointer accordingly upscale or down scale.

7   After the measurement is completed, return the pointer to zero, using the red surge-adjustment screw. Recheck the zero setting, using the black, zero-adjust screw.

How To Take Voltage Readings

Voltage readings are always taken across the circuit (between two phases or one phase to ground) by means of test leads. On the instrument under discussion, the test leads are inserted into the voltage receptacles at the bottom of the instrument. The rotary scale selector is then turned until its highest voltage range, usually 600 volts, appears in the scale window.

Connect one alligator clip to one side of the line. Then, with meter in hand, touch the other side of the line with the alligator clip. If the voltage does not exceed 600 volts, attach the second alligator clip and read the voltage on the red scale marked 600 volts. If the voltage does not exceed 300 volts, attach the second alligator clip and read the voltage on the red scale marked 300 volts, as shown in Figure 17–5. If voltage is below 150, rotate the scale selector until the 150-volt range appears in the window. Read the voltage on this scale.

To show how to read the scale, let us assume that the pointer is at the position indicated in the scales in Figure 17–6. In Figure 17–6a, the pointer reads 440 volts; each subdivision between 400 and 500 is 20 volts. In Figure 17–6b, the pointer reads 78 amperes; each subdivision between 70 and 80 is 2 amperes. The heavy mark between 60 and 80 is 70 amperes. Figure 17–6c shows a reading of 12.7 amperes. The heavy mark above 12 is 13 amperes, and each subdivision between 12 and 13 is 0.5 amperes.

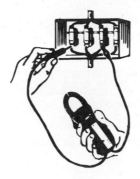

Figure 17–5   METHOD OF TAKING VOLTAGE READING BETWEEN PHASES WITH A VOLTMETER.

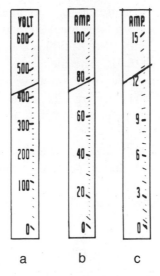

a   b   c

Figure 17–6a   600-VOLT SCALE; POINTER IS AT 440 VOLTS.
Figure 17–6b   100-AMPERE SCALE; POINTER READS 78 AMPERES.
Figure 17–6c   15-AMPERE SCALE; POINTER READS 12.7 AMPERES.

How To Use an Ohmmeter

The instrument under discussion may be changed to operate as an ohmmeter by inserting one of the test leads into the left (with the scale facing you) voltage receptacle at the bottom of the instrument. Insert the other test lead into the bottom of the separate battery case, then push in and lock in place. The opposite end of the battery case is then plugged into the jack on the right-hand side of the instrument just below an ohmmeter zero-adjustment knob.

Set the range selector so that the 150-volt, red scale appears in the scale window. The ohmmeter scale should be adjusted with the test leads open or not touching each other. In this position, the pointer should line up with the division marked ∝ (infinity) on the ohms scale. If it does not, turn the pointer zero-adjust screw until it does line up properly.

Now touch the test leads together and the pointer should jump to zero on the scale. If not, use the scale-adjustment knob to line the pointer to zero on the scale.

To measure resistance, make certain that no voltage is present on the device under test and then place the test leads between any two points on which resistance reading is desired. The ohmmeter scale is located on the flat plate to the right of the window. The zero mark (beginning) is on top of the scale, while the infinity mark ∝ ends the scale.

17–2   MEGGER

A typical megger (Figure 17–7) is composed of a hand-driven a.c. generator and/or a transformer with voltage rectified to 100, 250, 500 and 1000 volts d.c.; a cross-coil movement with 0 to 20,000 ohms and 0 to 1000 megohms scales; a carrying case; and test leads. The megger is used to measure the resistance, in megohms, to the flow of current through and/or over the surface of electrical equipment insulation. The test results are used to detect the presence of dirt, moisture, and insulation deterioration. The instrument also measures resistances up to 20,000 ohms.

The test set and the sample to which it is connected are sources of high-voltage electrical energy, and all persons making or assisting in the tests must use all practical safety precautions to prevent contact with energized parts of the test equipment and associated circuits. Persons actually engaged in the test must stand clear of all parts of the complete high-voltage circuit unless the set is deenergized and all parts of the test circuit are grounded. If the set is properly operated and all grounds are correctly made, no rubber gloves are necessary. As a routine safety procedure, however, some users require the use of rubber gloves in making connections to the high-voltage terminals and in manipulating the controls.

Figure 17–7   A POPULAR MODEL MEGGER.

The instruction manuals accompanying the megger contain detailed instructions about preparing for tests and connecting the megger to various types of equipment. Figures 17–8, 17–9, and 17–10 which give some practical points on hook ups of megger instruments are furnished by James G. Biddle, Co.

Figure 17–8 shows that a.c. motors and starting equipment can be megger-tested by connecting one side of the megger to the motor side of the main switch and the second test connection to a clip on the motor housing. If an insulation weakness is indicated by this test, the motor and starter should be checked separately.

Figure 17–9 shows the connection for testing low-voltage power cable. After both ends of the cable have been disconnected, the conductors are tested, one at a time, by connecting one of the megger leads to the conductor under test, and connecting the remaining conductors (within the cable) to ground and then to the other (ground) test lead. Similarly other insulation resistances, such as between conductor and outside protective sheath and between conductors, can be measured.

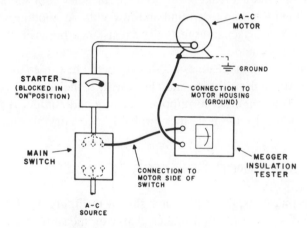

Figure 17–8 MEGGER CONNECTIONS FOR TESTING A.C. MOTOR AND STARTING EQUIPMENT.

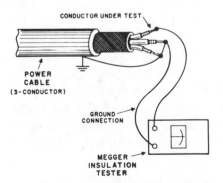

Figure 17–9 MEGGER FOR TESTING LOW-VOLTAGE POWER CABLE.

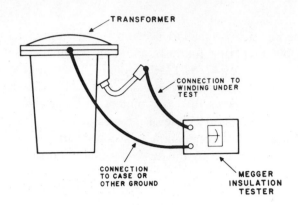

Figure 17–10   MEGGER CONNECTION FOR TESTING TRANSFORMERS.

Lighting and distribution transformers are tested by, first, making sure that switches or circuit breakers on both the primary and secondary sides are open. High-voltage winding to ground and low-voltage winding to ground are consecutively checked by separate tests. The resistance between the two is then checked with neither grounded.

If an insulation weakness is shown by the test in an oil-filled transformer, the dielectric strength of the oil may be tested; if the reading is low, the transformer should be filled in order to remove moisture and sludge. If the windings still give a poor reading by the test, they probably need drying out. See Figure 17–10.

17–3   PHASE-SEQUENCE INDICATOR

The phase-sequence indicator shown in Figure 17–11 is designed for use in conjunction with any multimeter that can measure a.c. voltage. This phase-sequence adapter can be used on circuits with line voltages up to 550 volts a.c., provided the instrument used with the indicator has a rating this high.

To use this indicator, set the multimeter to the proper voltage range. This can be determined (if it is not known) by measuring the line voltage before connecting the phase-sequence indicator. Next, connect the two black leads of the indicator to the voltage test leads of the meter, as shown in Figure 17–12. Connect the red, yellow, and black adaptor leads to the circuit in any order and check the meter for voltage reading. If the meter reading is higher than the original circuit voltage measured, then the phase sequence is Black-Yellow-Red. If the meter reading is lower than the original circuit voltage measured, then the phase sequence is Red-Yellow-Black. If the reading is the same as the first reading, then one phase is open.

Figure 17–11   AN ADAPTOR FOR TESTING PHASE SEQUENCE.

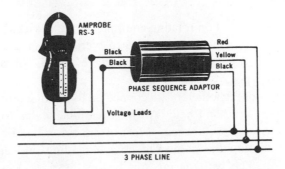

Figure 17–12   CONNECTIONS FOR PHASE-SEQUENCE INDICATOR.

17–4   POWER-FACTOR METER

The power factor of an electrical circuit or system is the ratio of true power to the apparent power. Power factor depends on the phase difference between the current in and the voltage across an a.c. circuit. The power factor (P.F.) of a circuit may be obtained (by calculation) by taking simultaneous readings with an ammeter, a voltmeter, and a wattmeter and dividing the watts by the product of the voltage and the current:

$$\text{P.F. (single-phase circuit)} = \frac{\text{watts}}{\text{volts} \times \text{amperes}}$$

Much time can be saved by using a single instrument to determine the power factor of a given circuit. A power factor meter indicates directly the factor values without need for any further calculations. Such meters are constructed so that the rotating field is produced by the line voltage- and there is no torque on the moving coil at a power factor of 100 percent or unity, so the pointer rests at the center of the scale, normally designated ''100.'' Depending on whether the current is leading or lagging when the power factor is less than unity, the pointer swings to the left or to the right on the scale, indicating directly the value of the power factor for the circuit being tested.

Three-phase, power-factor meters are available and are commonly used in industrial applications. Most industrial plants are charged a penalty by the power company if the power factor of their overall electrical system falls below 90 percent. Therefore, industries try to maintain a high power factor at all times, and the power-factor meter is a great aid in doing so.

17–5 TACHOMETER

Tachometers, used on rotating equipment, are devices that indicate or record the speed of a rotating motor or machine. Several types are available, but the most popular are centrifugal, eddy current, surface speed, vibrating reed, and high-intensity photo types.

Some types of tachometers are connected to the machine by means of belts or gears, but others are hand held. All have various scales from 0 revolutions per minute (rpm) to over 50,000 rpm. Most of the better hand-held models have two buttons—one to stop the pointer to hold a reading and another to release the pointer to zero. The ball-bearing spindle on these devices is placed against a rotating object and the speed of the shaft, wheel, etc., is read directly on the dial of the tachometer.

Totally enclosed rotating equipment may be checked for speed by using the vibrating-reed tachometer that operates on the well known and time-tested principle of resonance. The instrument is simply held against the motor, turbine, pump, compressor, or other rotating equipment and the speed is shown by the vibrating-reed tachometer, which consists of a steel reed tuned to a certain standard speed.

The light of the photo-type tachometer is aimed at the rotating shaft on which there is a contrasting color such as a mark, chalk line, or a light-reflective strip or tape. (Did you ever watch the timing being set on your automobile engine?) The rotational speed in rpm is immediately read directly from the indicating scale of the instrument. This type of tachometer is especially useful on relatively inaccessbile rotational equipment such as motors, fans, and grinding wheels.

An electrical tachometer consists of a smaller a.c. or d.c. generator belted or geared to the unit under test. The voltage produced in the generator varies directly with the speed of the rotating part of the generator. Since this speed is directly proportional to the speed of the machine under test, the amount of the generated voltage is a measure of the speed. The generator is electrically connected to an indicating or recording instrument that is calibrated to indicate units of speed such as rpm, fps, and fpm.

17–6 LIGHT-INTENSITY METER

There are a number of large and elaborate devices used in laboratories for making exact tests and measurement of lamps and lighting fixtures; however, for practical use in the field, a portable *footcandle* meter, as shown in Figure 17–13,

Figure 17–13 TYPICAL LIGHT-INTENSITY METER.

is quite satisfactory. This instrument consists of light barrier layer cells and a meter enclosed in a suitable covering and reads light intensities from 1 to 500 or more footcandles.

To use a footcandle meter of this type, remove the cover and hold the meter in such a position that the cell is facing towards the light source—at the level of the work plane where the illumination is required. Your shadow should not fall on the cell during the test. A number of such tests at various points in a room or area will give the average illumination level in footcandles.

17–7 FREQUENCY METER

Modern a.c. systems must be closely regulated to a constant (usually 60-Hz) frequency—normally to within 1 percent of their rating. To measure these small variations, very sensitive frequency meters are required. Most are either of the ratio-meter type or the vibrating-reed type. The latter type contains metal reeds that are energized by the electromagnetic field. When the frequency of the circuit under test is the same as the frequency of one of the reeds, the reed vibrates so that the vibration is visibly evident. The scale opposite each reed is marked with the circuit frequency indicated.

Since the frequency of an alternator depends on the number of poles it has and on the speed of rotation of its rotating part, the scale of a frequency meter may be marked to indicate the number of revolutions per minute of the alternator. The frequency meter can then be used as a speed indicator by merely placing the meter to the alternator; it will pick up enough vibrations to give an indication of the speed without any electrical connection.

17–8 ELECTRICAL THERMOMETERS

To measure the temperature of various items or areas, three basic electrical methods are used: the resistance method, the thermocouple method, and the radiation-, optical-pyrometer methods. The resistance method makes use of the fact that the resistance of a metal varies in direct proportion to its temperature. This method is normally used to measure temperatures up to approximately 1500°F (800°C).

The thermocouple method is based on the principle that a difference in temperature in different metals generates a voltage. This method is used to measure temperatures up to about 3000°F (1650°C).

The radiation-pyrometer and the optical-pyrometer are generally used for temperatures above 3000°F (1650°C). These instruments combine the principle of the thermocouple with the effect of radiation of heat and light.

QUESTIONS

Answer the following questions by filling in the blanks.

1   When a resistance test is taken with an ohmmeter, make certain that _____ voltage is present on the device or circuit under test.

2   Of the several methods used to measure temperature, the _____ or _____ methods would normally be used to measure the temperature of a furnace whose temperature is suspected to be around 5800°F.

3   Besides measuring resistance, the megger is used to detect _____, _____, and _____ in an electrical circuit or in a piece of equipment.

4   In a phase-sequence test, when an adaptor is used in conjunction with a multimeter, a meter reading higher than the original circuit voltage indicates that the phase-sequence colors are _____, _____, and _____.

5   The power factor of an electrical system is the ratio of true power to the _____ power.

6   Most industrial plants are charged a penalty by the power company if the power factor of their electrical system falls below _____ percent.

7   _____ are devices used to indicate the speed of a rotating machine or motor.

8   Most light-intensity meters indicate the amount of illumination directly in _____.

9   Besides measuring variations in frequency, frequency meters also give an indication of the _____ of a motor or alternator.

10   A thermometer based on the fact that the resistance of a metal varies in direct proportion to its temperature uses what is known as the _____ method to indicate temperature.

Appendix
GLOSSARY

Accessible (As applied to wiring methods) Capable of being removed or exposed without damaging the building structure or finish or not permanently closed in by the structure or finish of the building.

Accessible (As applied to equipment) Admitting close approach because not guarded by locked doors, elevation, or other effective means.

Aggregate Inert material mixed with cement and water to produce concrete.

Alternating current (a.c.) Alternating current is current that reverses direction rapidly, flowing back and forth in the system with regularity. This reversal of current is due to reversal of voltage, which occurs at the same frequency. In alternating current, any one wire is first positive, then negative, then positive, and so on.

Alternator An electric generator designed to supply alternating current. Some types have a revolving armature, and other types have a revolving field.

Ampacity Current carrying capacity, expressed in amperes.

Ampere The unit of measurement for electric current. It represents the rate at which current flows through a resistance of one ohm by a pressure of one volt.

Amplitude The maximum instantaneous value of an alternating voltage or current. It is measured in either the positive or negative direction.

Appliance Utilization equipment, generally equipment other than industrial, normally built in standardized sizes or types and installed or connected as a unit to perform one or more functions, such as clothes washing, air conditioning, food mixing, or deep frying.

Appliance, fixed An appliance that is fastened or otherwise secured at a specific location.

Appliance, portable An appliance that is actually moved or can easily be moved from one place to another in normal use.

Appliance, stationary   An appliance that is not easily moved from one place to another in normal use.

Approved   Acceptable to the authority enforcing the Code.

Attachment plug   (plug cap) (cap) A device that upon insertion in a receptacle establishes a connection between the conductors of the attached flexible cord and the conductors connected permanently to the receptacle.

Automatic   Self-acting, operating by its own mechanism when actuated by some impersonal influence, such as a change in current strength, pressure, temperature, or mechanical configuration.

Backfill   Loose earth placed outside foundation walls for filling and grading.

Bearing plate   Steel plate placed under one end of a beam or truss for load distribution.

Bearing wall   Wall supporting a load other than its own weight.

Bench mark   Point of reference from which measurements are made.

Bonding jumper   A reliable conductor used to ensure the required electrical conductivity between metal parts required to be electrically connected.

Branch circuit   That portion of a wiring system extending beyond the final overcurrent device protecting the circuit.

Branch circuit, appliance   A circuit supplying energy to one or more outlets to which appliances are to be connected. Such circuits have no permanently connected lighting fixtures that are not a part of an appliance.

Branch circuit, general purpose   A branch circuit that supplies a number of outlets for lighting and appliances.

Branch circuit, individual   A branch circuit that supplies only one piece of utilization equipment.

Bridging   System of bracing between floor beams to distribute the floor load.

Building   A structure that stands alone or that is cut off from adjoining structures by fire walls with all openings therein protected by approved fire doors.

Bus bar   The heavy copper or aluminum bar used on switch-boards to carry current.

Cabinet   An enclosure designed for either surface or flush mounting and provided with a frame, mat, or trim in which swinging doors are hung.

Capacitor or condenser   An electrical device that causes the current to lead the voltage, opposite in effect to inductive reactance. It is used to neutralize the objectional effect of lagging (inductive reactance), which overloads the power source. It also acts as a low resistance path to ground for currents of radio frequency, thus effectively reducing radio disturbance.

Cavity wall   Wall built of solid masonry units arranged to provide air space within the wall.

Chase   Recess in inner face of masonry wall providing space for pipes and/or ducts.

Circuit breaker   A device designed to open and close a circuit by nonautomatic means and to open the circuit automatically on a predetermined overload of current, without injury to itself when properly applied within its rating.

Coaxial cable   A cable consisting of two conductors concentric with and insulated from each other.

Column   Vertical load-carrying member of a structural frame.

Commutator   Device used on electric motors or generators to maintain a unidirectional current.

Concealed   Rendered inaccessible by the structure or finish of the building. Wires in concealed raceways are considered concealed even though they may become accessible by withdrawing them.

Conductance   The ability of material to carry an electric current.

Conductor   Substances that offer little resistance to the flow of electric current. Silver, copper, and aluminum are good conductors, but no material is a perfect conductor.

Conductor, bare   A conductor having no covering or insulation whatsoever.

Conductor, covered   A conductor having one or more layers of nonconducting materials that are not recognized as insulation under the NEC.

Conductor insulated   A conductor covered with material recognized as insulation.

Connector, pressure   (solderless) A connector that establishes the connection between two or more conductors or between one or more conductors and a terminal by means of mechanical pressure and without the use of solder.

Continuous load   A load in which the maximum current is expected to continue for three hours or more.

Contour line   A line connecting points with the same elevation on a map denoting elevations.

Controller   A device or group of devices that serves to govern in some predetermined manner the electric power delivered to the apparatus to which it is connected.

Cooking unit, counter-mounted   An assembly of one or more domestic surface heating elements for cooking purposes designed to be flush mounted in, or supported by, a counter and complete with internal wiring and inherent or separately mounted controls.

Crawl space   Shallow space between the first tier of beams and the ground (no basement).

Current   The flow of electricity is a circuit. It is expressed in amperes and represents an amount of electricity.

Curtain wall   Nonbearing wall between piers or columns for the enclosure of the structure, not supported at each story.

Cycle   One complete period of flow of alternating current in both directions. One cycle represents 360 degrees.

Demand factor   In any system or part of a system, the ratio of the maximum demand of the system, or part of the system, to the total connected load of the system, or part of the system under consideration.

Direct current (d.c.)   Current that flows in one cirection only with one terminal always positive and the other always negative.

Disconnecting means   A device, group of devices, or other means whereby the conductors of a circuit can be disconnected from their source of supply.

Double-strength glass   Sheet glass that is 1/8-inch thick (single-strength glass is 1/10-inch thick.)

Dry wall   Interior wall construction consisting of plaster boards, wood paneling, or plywood nailed directly to the studs without application of plaster.

Duty, continuous   A requirement of service that demands operation at a substantially constant load for an indefinitely long time.

Duty, intermittent   A requirement of service that demands operation for alternate intervals of load and no load; load and rest; or load, no load, and rest.

Dry, periodic   A type of intermittent duty in which the load conditions regularly recur.

Duty, short-time   A requirement of service that demands operations at loads and for intervals of time, both of which may be subject to wide variation.

Electrical generator   An electrical generator is a machine so constructed that a voltage is generated when its rotor is driven by an engine or other prime mover.

Electrode   A conducting element used to emit, collect, or control electrons and ions.

Electron   A particle of matter negatively charged.

Electron emission   The release of electrons from the surface of a material into surrounding space due to heat, light, high voltage, or other causes.

Elevation   Drawing showing the projection of a building on a vertical plane.

Enclosed   Surrounded by a case that prevents anyone from accidentally contacting live parts.

Equipment   A general term including material, fittings, devices, appliances, fixtures, and apparatus used as a part of, or in connection with, an electrical installation.

Expansion bolt   Bolt with a casing arranged to wedge the bolt into a masonry wall to provide an anchorage.

Expansion joint   Joint between two adjoining concrete members arranged to permit expansion and contraction with changes in temperature.

Exposed   (As applied to live parts) Live parts that a person could inadvertently touch or approach closer than a safe distance. This term is applied to parts that are not suitably guarded, isolated, or insulated.

Exposed   (As applied to wiring) Not concealed.

Externally operable   Capable of being operated without exposing the operator to contact with live parts.

Facade   Main front of a building.

Farad   A measure of electrical capacity of condensers.

Feedback   The process of transferring energy from the output circuit of a device back to its input.

Feeder   The conductors between the service equipment or the generator switchboard of an isolated plant and the branch-circuit overcurrent device.

Frequency   Frequency alternating current is the number of cycles per second. A 60-hertz alternating current makes 60 complete cycles of flow back and forth (120 alternations) per second. A conventional alternator has an even number of field poles arranged in alternate north and south polarities. Current flows in one direction in an a.c. armature conductor when the conductor is passing a north pole and in the other direction when it is passing a south pole. The conductor passes two poles during each cycle. A frequency of 60 hertz requires the conductor to pass 120 poles per second. In a six-pole alternator, the equivalent speed would be 30 revolutions per second or 1800 revolutions per minute.

Filter   A combination of circuit elements designed to pass a definite range of frequencies, reducing all others.

Fire stop   Incombustible filler material used to block interior draft spaces.

Fitting   An accessory, such as a locknut, bushing, or other part of a wiring system, that is intended primarily to perform a mechanical rather than an electrical function.

Flashing   Strips of sheet metal bent into an angle between the roof and wall to make a water-tight joint.

Footing   Structural unit used to distribute loads to the bearing materials.

Frostline   Deepest level in a geographic area below grade to which frost penetrates.

Fuse   A protective device inserted in series with a circuit.

Gain   The ratio of output to input power, voltage, or current.

Garage   A building or portion of a building in which one or more self-propelled vehicles carrying volatile, flammable liquid for fuel or power are kept for use, sale, storage, rental, repair, etc.

Ground   A conducting connection, intentional or accidental, between an electrical circuit or piece of equipment and earth or some other conducting body serving in place of the earth.

Grounded   Connected to earth or to some conducting body that serves in place of the earth.

Grounded conductor   A system or circuit conductor that is intentionally grounded.

Grounding conductor   A conductor used to connect equipment or the grounded circuit of a wiring system to a grounding electrode.

Henry   The basic unit of inductance.

Hertz   A unit of frequency, one cycle per second. Written as 50-hertz or 60-hertz current.

I beam   Rolled steel beam or built-up beam of I section.

Impedance   Effects placed on alternating current by inductive capacitance (current lags voltage), capacitive reactance (current leads voltage) and resistance (opposes current but does not lag or lead voltage), or any combination of the two. It is measured in ohms, like resistance.

Incombustible material   Material that will not ignite or actively support combustion in a surrounding temperature of 1200°F during an exposure of 5 minutes; also material that will not melt when the temperature of the material is maintained at 900°F for a period of 5 minutes.

Inductance   The property of a circuit or two neighboring circuits that determines how much voltage will be induced in one circuit by a change of current in either circuit.

Inductor   A coil.

Integrated circuit   A circuit in which different types of devices, such as resistors, capacitors, and transistors, are made from a single piece of material and are connected to form a circuit.

Insulator   Substances that offer a greater resistance to the flow of electric current, such as glass, porcelain, paper, cotton, enamel, and paraffin, are called insulators because they are practically nonconducting. However, no material is a perfect insulator.

Isolated   Not readily accessible to persons unless special means for access are used.

Jamb   Upright member forming the side of a door or window opening.

kVA   The abbreviation of kilovolt-amperes, which is the product of volts times amperes divided by 1000. This term is used in rating alternating-current machinery because the product of volts and amperes with alternating currents usually does not give the true average power.

kVAR   The abbreviation of kilovolt-ampere reactance, which is a measurement of reactive power that generates power within induction equipment (motors, transformers, holding coils, lighting ballasts, etc.).

kW   The abbreviation of kilowatt, which is a unit of measurement of electrical power. A kilowatt (kW) equals 1000 watts and is the product of volts times amperes divided by 1000, when used in rating direct current machinery. The term is used to indicate true power in an a.c. circuit.

Kilowatt hour   A kilowatt hour is the amount of electrical power represented by 1000 watts for a period of 1 hour. Thus, a generator that delivers 1000 watts for a period of 1 hour would have delivered 1 kilowatt hour of electricity.

Lally column   Compression member consisting of a steel pipe filled with concrete under pressure.

Laminated wood   Wood build up of plies or laminations that have been joined either with glue or with mechanical fasteners. Usually, the plies are too thick to be classified as veneer, and the grain of all plies is parallel.

Lighting outlet   An outlet intended for the direct connection of a lampholder, lighting fixture, or pendant cord terminating in a lampholder.

Location, damp   A location subject to a moderate amount of moisture, such as some basements, barns, and cold-storage warehouses.

Location, dry   A location not normally subject to dampness or wetness; a location classified as dry may be temporarily subject to dampness or wetness, as in the case of a building under construction.

Location, wet   A location subject to saturation with water or other liquids, such as locations exposed to weather and washrooms in garages. Installations that are located underground, in concrete slabs, or in masonry in direct contact with the earth shall be considered wet locations.

Logic   The arrangement of circuitry designed to accomplish certain objectives.

Low energy-power circuit   A circuit that is not a remote-control or signal circuit but whose power supply is limited in accordance with the requirements of Class-2, remote-control circuits.

Modulation   The process of varying the amplitude, frequency, or the phase of a carrier wave.

Multioutlet assembly   A type of surface or flush raceway designed to hold conductors and attachment plug receptacles that can be assembled in the field or at the factory.

National Electrical Code   The National Electrical Code is sponsored by the National Fire Protection Association and is the ''Bible'' of all electrical workers for building construction. It is often referred to as the ''NEC'' or the ''Code.''

Nonautomatic   An action requiring personal intervention for its control.

Ohm   The ohm is the unit of measurement of electrical resistance and represents the amount of resistance that permits current to flow at the rate of one ampere under a pressure of one volt. The resistance (in ohms) equals the pressure (in volts) divided by the current (in amperes).

Outlet   In the wiring system, a point at which current is taken to supply utilization equipment.

Outline lighting   An arrangement of incandescent lamps or gaseous tubes to outline and call attention to certain features, such as the shape of a building or the decoration of a window.

Oven, wall-mounted   A domestic oven for cooking purposes designed for mounting into or onto a wall or other surface.

Panelboard   A single panel or group of panel units designed for assembly in the form of a single panel; includes buses and may come with or without switches and/or automatic overcurrent protection devices for the control of light, heat, or power circuits of small individual as well as aggregate capacity. It is designed to be placed in a cabinet or cutout box placed in or against a wall or partition and to be accessible only from the front.

Pilaster   Flat square column attached to a wall and projecting about a fifth of its width from the face of the wall.

Plate   The principal anode in an electron tube to which the electron stream is attracted.

Plenum   Chamber or space forming a part of an air-conditioning system.

Potential   The difference in voltage between two points of circuit. Frequently one is assumed to be ground (zero potential).

Potentiometer   An instrument for measuring an unknown voltage or potential difference by balancing it, wholly or in part, by a known potential difference produced by the flow of known currents in a network of circuits of known electrical constants.

Power   The rate of doing work or expending energy.

Power factor   When the current waves in an alternating-current circuit coincide exactly in time with the voltage waves, the product of volts times amperes gives volt amperes which is true power in watts (or in kW if divided by 1000). When the current waves lag behind the voltage due to inductive reactance (or lead due to capacitive reactance), they do not reach their respective values at the same time. Under such conditions, the product of volts and amperes does not give true average watts. Such a product is called volt amperes or apparent watts. The factor by which apparent watts must be multiplied to give the true watts is known as the power factor (P.F.). The power factor depends on the amount of lag or lead and is the percentage of apparent watts that represents true watts.

With a power factor of 80 percent, a fully loaded 5 kVA alternator will produce

4KW. When the rating of a power unit is stated in kVA at 80 percent P.F., it means that the generator with an 80 percent P.F. load will generate its rated voltage, providing the load does not exceed the kVA rating.

An engine-driven alternator with automatic voltage regulation, for example, the kVA rating is usually determined by the maximum current that can flow through the windings without injurious overheating or by the ability of the engine or other prime mover to maintain the normal operating speed. A resistance load, such as electric lamp bulbs, irons, and toasters, is a unity power factor load. Motors, transformers, and various other devices cause a current wave lag, which is expressed in the power factor of the load.

Precast concrete Concrete units (such as piles or vaults) cast away from the construction site and set in place.

Qualified person One familiar with the construction and operation of the apparatus and the hazards involved.

Raceway Any channel designed expressly for holding wire, cables, or bus bars and used solely for this purpose.

Rain tight So constructed or protected that exposure to a beating rain will not result in the entrance of water.

Reactance Reactance is opposition to the change of current flow in an a.c. circuit. The rapid reversing of alternating current tends to induce voltages that oppose the flow of current in such a manner that the current waves do not coincide in time with the voltage waves. The opposition of self-inductance to the flow of current is called inductive reactance and causes the current to lag behind the voltage that produces it. The opposition of a condenser or of capacitance to the change of a.c. voltage causes the current wave to lead the voltage wave. This is called capacitive reactance. The unit of measurement for either inductive reactance or capacitive reactance is the ohm.

Readily accessible Capable of being reached quickly for operation or inspection without requiring those to whom ready access is requisite to climb over or remove obstacles or to resort to portable ladders, chairs, etc.

Receptacle (convenience outlet) A contact device installed at an outlet for the connection of an attachment plug.

Receptacle outlet An outlet where one or more receptacles are installed.

Rectifiers Devices used to change alternating current to unidirectional current.

Relay An electromechanical switching device that can be used as a remote control.

Remote-control circuit Any electrical circuit that controls any other circuit through a relay or an equivalent device.

Resistance Electrical resistance is opposition to the flow of electric current and may be compared with the resistance of a pipe to the flow of water. All substances have some resistance, but the amount varies with different substances and with the same substances under different conditions.

Resistor A resistor is a poor conductor used in a circuit to create resistance that limits the amount of current flow. It may be compared with a valve in a water system.

Resonance   In a circuit containing both inductance and capacitance, resonance is a condition in which the inductive reactance is equal to and cancels out the capacitance reactance.

Riser   Upright member of stair extending from tread to tread.

Roughing in   Installation of all concealed electrical wiring; includes all electrical work done before finishing.

Saturation   The condition existing in a circuit when an increase in the driving signal does not produce any further change in the resultant effect.

Sealed (hermetic-type) motor compressor   A mechanical compressor consisting of a compressor and a motor, both of which are enclosed in the same sealed housing, with no external shaft or shaft seals, the motor operating in the refrigerant atmosphere.

Semiconductor   A material midway between a conductor and an insulator.

Service   The conductors and equipment used for delivering energy from the electricity supply system to the wiring system of the premises served.

Service cable   The service conductors made up in the form of a cable.

Service conductors   The supply conductors that extend from the street main or transformers to the service equipment of the premises being supplied.

Service drop   The overhead service conductors from the last pole or other aerial support to and including the splices, if any, that connect to the service-entrance conductors at the building or other structure.

Service-entrance conductors, underground system   The service conductors between the terminals of the service equipment and the point of connection to the service equipment and the point of connection to the service lateral.

Service equipment   The necessary equipment, usually consisting of a circuit breaker or switch and fuses and their accessories, located near the point of entrance of supply conductors to a building and intended to constitute the main control and means of cutoff for the supply to that building.

Service lateral   The underground service conductors between the street main, including any risers at a pole or other structure or from transformers, and the first point of connection to the service-entrance conductors in a terminal box, meter, or other enclosure with adequate space, inside or outside the building wall. Where there is no terminal box, meter, or other enclosure with adequate space, the point of connection shall be considered to be the point of entrance of the service conductors into the building.

Service raceway   The rigid metal conduit, electrical metallic tubing, or other raceway that encloses the service-entrance conductors.

Setting   (of circuit breaker) The value of the circuit at which the circuit breaker is set to trip.

Sheathing   First covering of boards or paneling nailed to the outside of the wood studs of a frame building.

Siding   Finishing material that is nailed to the sheathing of a wood frame building and that forms the exposed surface.

Signal circuit   Any electrical circuit supplying energy to an appliance that gives a recognizable signal.

Single phase   A single-phase, a.c. system has a single voltage in which voltage reversals occur at the same time and are of the same alternating polarity throughout the system.

Soffit   Underside of a stair, arch, or cornice.

Solenoid   An electromagnet having a movable iron core.

Soleplate   Horizontal bottom member of wood-stud partition.

Studs   Vertically set skeleton members of a partition or wall to which lath is nailed.

Switch, general-use   A form of general-use switch so constructed that it can be installed in flush device boxes or on outlet covers, or can otherwise be used in conjunction with wiring systems recognized by this Code.

Switch, a.c. general-use snap   A form of general-use snap switch suitable only for use on a.c. circuits and for controlling the following:

1   Resistive and inductive loads (including electric discharge lamps) not exceeding the ampere rating at the voltage involved.

2   Tungsten-filament lamp loads not exceeding the ampere rating at 120 volts.

3   Motor loads not exceeding 80 percent of the ampere rating of the switches at the rated voltage.

Switch, a.c.-d.c. general-use snap   A form of general-use snap switch suitable for use on either a.c. or d.c. circuits and for controlling the following:

1   Resistive loads not exceeding the ampere rating at the voltage involved.

2   Inductive loads not exceeding one-half the ampere rating at the voltage involved, except that switches having a marked horsepower rating are suitable for controlling motors not exceeding the horsepower rating of the switch at the voltage involved.

3   Tungsten-filament lamp loads not exceeding the ampere rating at 125 volts, when marked with the letter T.

Switch, isolating   A switch intended for isolating an electric circuit from the source of power. It has no interrupting rating and is intended to be operated only after the circuit has been opened by some other means.

Switch, motor-circuit   A switch, rated in horsepower, capable of interrupting the maximum operating overload current of a motor having the same horsepower rating as the switch at the rated voltage.

Switchboard   A large single panel, frame, or assembly of panels, having switches, overcurrent and other protective devices, buses, and usually instruments mounted on the face and/or back. Switchboards are generally accessible from the rear as well as from the front and are not intended to be installed in cabinets.

Synchronous   Simultaneous in action and in time (in phase).

Tachometer   An instrument for measuring revolutions per minute.

Thermal protector   (as applied to motors) A protective device that is assembled as an integral part of a motor or motor compressor and that, when properly applied, protects the motor against dangerous overheating due to overload and failure to start.

Thermalcutout   An overcurrent protection device containing a heater element in addition to and affecting a renewable fusible member that opens the circuit. It is not designed to interrupt short-circuit currents.

Thermally protected   (as applied to motors) Refers to these words appearing on the nameplate of a motor or motor compressor and means that the motor is provided with a thermal protector.

Three phase   A three-phase, a.c. system has three individual circuits or phases. Each phase is timed so the current alternations of the first phase is 1/3 cycle (120°) ahead of the second and 2/3 cycle (240°) ahead of the third.

Transformer   A device used to transfer energy from one circuit to another. It is composed of two or more coils linked by magnetic lines of force.

Trusses   Framed structural pieces consisting of triangles in a single plan for supporting loads over spans.

Utilization equipment   Equipment that utilizes electric energy for mechanical, chemical, heating, lighting, or similar useful purposes.

Ventilated   Provided with a means to permit circulation of air sufficient to remove an excess of heat, fumes, or vapors.

Volt   The practical unit of voltage or electromotive force. One volt sends a current of one ampere through a resistance of one ohm.

Voltage   The force, pressure, or electromotive force (EMF) that causes electric current to flow in an electric circuit. Its unit of measurement is the volt. Voltage in an electric circuit may be considered to be similar to water pressure in a pipe or water system.

Voltage drop   The voltage drop in an electrical circuit is the difference between the voltage at the power source and the voltage at the point at which electricity is to be used. The voltage drop or loss is created by the resistance of the connecting conductors.

Voltage to ground   In grounded circuits, the voltage between the given conductor and that point or conductor of the circuit that is grounded; in ungrounded circuits, the greatest voltage between the given conductor and any other conductor of the circuit.

Water tight   So constructed that moisture will not enter the enclosing case or housing.

Watt   The watt is the unit of measurement of electrical power or rate of work. 746 watts is equivalent to 1 horse-power. The watt represents the rate at which power is expended when a pressure on one volt causes current to flow at the rate of one ampere. In a d.c. circuit or in an a.c. circuit at unity (100 percent) power factor, the number of watts equals the pressure (in volts) multiplied by the current (in amperes).

Weatherproof   So constructed or protected that exposure to the weather will not interfere with successful operation.

Web   Central portion of an I beam.

INDEX